Recent Results in Cancer Research

Fortschritte der Krebsforschung

Progrès dans les recherches sur le cancer

48

Platinum Coordination Complexes in Cancer Chemotherapy

Edited by

T. A. Connors and J. J. Roberts

With 93 Figures

Springer-Verlag New York · Heidelberg · Berlin 1974

Second International Symposium on Platinum Coordination Complexes in Cancer Chemotherapy, Oxford, April 1973

Sponsored by the Swiss League against Cancer

Library of Congress Cataloging in Publication Data

International Symposium on Platinum Coordination Complexes in Cancer Chemotherapy, 2d, Oxford University, 1973.
Platinum coordination complexes in cancer chemotherapy.
(Recent results in cancer research, 48)
1. Cancer—Chemotherapy—Congresses. 2. Organoplatinum compounds—Therapeutic uses—Congresses. I. Connors, T. A., 1934— ed. II. Roberts, J. J., 1926— ed. III. Title. IV. Series. [DNLM: 1. Antineaplastic agents. 2. Diammonium chloroplatinum compounds—Therapeutic use. 3. Neoplasms—Drug therapy. 4. Platinum—Therapeutic use.
W1 RE106P v. 48 / QZ267 P716] RC261.R35 no. 48 [RC271.P55] 616.9'94'008s 74-12406
[616.9'94'061]

ISBN 978-3-642-49308-9 ISBN 978-3-642-49306-5 (eBook)
DOI 10.1007/978-3-642-49306-5

Foreword

It was a great pleasure and honour to have been invited to attend this Seminar and to present a final impression. The association in this field between the Chester Beatty Research Institute and Dr. Rosenberg's School at East Lansing is something which I specially value and many would doubtless like to know how it came about.

In the course of its work in carcinogenesis and on chemotherapy over many years, the Chester Beatty Research Institute was frequently drawn to the importance of many metals — as for example lead, iron, metalloid arsenic and the metalloid qualities of the carcinogenic hydrocarbons. Interest started in platinum many years ago, following the possibility, claimed by others, that various complexes between the metal and mercaptopurine might possess significant chemotherapeutic properties. Various attempts to confirm such findings ended, however, in complete failure. Interest in platinum was revived by the fresh observations of Dr. Rosenberg and his colleagues, and here the outcome was entirely different. Very soon it was possible to confirm the intense growth-inhibitory properties of *cis*-platinum (II) diamminedichloride and related substances. After communicating these results to Dr. Rosenberg, it was a pleasure to welcome him in London where he gave a Seminar which greatly engaged the interest of many of the staff. Later, several of these were to enjoy Dr. Rosenberg's hospitality at an international conference on the subject to be held in East Lansing, where many rapidly developing aspects were open for discussion. Thereafter, co-operation between the two groups remained active.

Having some interest in the methods and philosophy of research as well as in its details, the manner of discovery of the platinum effect has never ceased to have its special fascination being a mixture of accident and great shrewdness. In many ways it has the attributes, say, of the detection of penicillin or that of bacteriophage. Our own interest rapidly mounted with the realisation of the alkylating-like properties of the platinum complexes, and aslo of the possibility that they might engage like classical alkylating agents in chemical cross linkage.

Listening to the accounts of clinical trial, mainly with *cis*-platinum (II) diamminedichloride it was inevitable that we should share impressions both discouraging and optimistic. I present no apologies for urging in the ultimate outcome an optimistic impression. We must remember that the progress of all such studies is inevitably slow, and that we are already aware of specific compounds vastly greater, with a chemotherapeutic index of say 200, than that of the compound which we are at the moment considering, which has an index of 8. And during these days we have learned of yet other compounds with chemical, physical and biological properties especially favourable. We therefore look forward to the future with some optimism, and I am constrained to do this from very wide experience of chemotherapeutic agents under very difficult circumstances over many years. But even so the subject in general still remains in its infancy, and its main growth has been largely confined to the post-war years.

I am reminded that while a medical student in the University of Edinburgh—admittedly a long time ago—one was taught not merely the principles of anti-protozoal chemotherapy, but also the reasons why no anti-bacterial chemotherapy could be expected. But as you know, the entire picture was revolutionised within a very few years. Contrarywise, one tends to be encouraged at the present time by a growing understanding of spontaneous regression, and by the chemotherapeutic responses, admittedly sometimes in concert with an immune component now being observed in Burkitt's Lymphoma, in choriocarcinoma and even in acute leukaemia. I am also impressed by the growing number of cases in which the regression of various animal tumours has been attributable to chemotherapy alone.

In these closing remarks, I would be failing in my duty were I not to acknowledge our special indebtedness to great generosity on the part of Dr. Rosenberg, to similar munificence on the part of industry and to the special arrangements made on our behalf by Wadham College and Dr. R. Williams who was an ideal host.

April 1974 ALEXANDER HADDOW

Contents

List of Participants

Carter, S. K., M.D., Cancer Therapy Evaluation Branch, Division of Cancer Treatment, National Cancer Institute, Bethesda, MD 20014/USA.

Cleare, M. J., Ph.D., Johnson Matthey and Co. Ltd., Wembley/England.

Connors, T. A., D.Sc., Chester Beatty Research Institute, Institute of Cancer Research, Royal Cancer Hospital, Fulham Road, London S. W. 3, J. B.

Conran, P. B., Ph.D., University of Connecticut, Health Center, Farmington, Ct 06032/USA.

Dienstbier, Z., M.D., Institute of Biophysics and Nuclear Medicine, Faculty of General Medicine, Charles University, Prague/CSSR.

Ellerby, R. A., M.D., Metropolitan Clinic, Physicians and Surgeons, 265 North Broadway, Portland, Oregon 97227/USA.

Harder, M. C., Ph.D., George Washington University, Medical Center, 2300 Eye Street, N. W., Washington DC 20037/USA.

Hill, J. M., M.D., Granville C. Morton Cancer and Research Hospital of the Wadley, Institute of Molecular Medicine, Dallas, TX/USA.

Hindy, I., M.D., National Oncological Institute, Rath Gyorgy UTCA 5, Budapest XII.

Krakoff, I. H., M.D., Medical Oncology Service, Department of Medicine, Memorial Hospital, Sloan-Kettering Institute, 444 East 68th Street, New York, NY, 10021/USA.

Roberts, J. J., D.Sc., Chester Beatty Research Institute, Institute of Cancer Research, Royal Cancer Hospital, London S. W. 3, 6 J. B., U. K.

Robins, A. B., Ph.D., Institute of Cancer Research, Sutton, Surrey/England.

Talley, R. W., M.D., Henry Ford Hospital, 2600 W Grand Boulevard, Detroit, Michigan/USA.

Thomson, A. J., Ph. D., School of Sciences, University of East Anglia, Norwich, Norfolk/England.

Wallace, H. J., M.D., Department of Medicine S, Roswell Park Memorial Institute, 666 Elm Street, Buffalo, NY 14203/USA.

Williams, R. J. P., D.Sc., Wadham College, Oxford/England.

Wiltshaw, E., M.D., Department of Chemotherapy, Royal Marsden Hospital, Institute of Cancer Research, Fulham Road, London S. W. 3.

Introduction

B. ROSENBERG

The medical use of metal complexes was unquestionably at its peak during the time of Paul Ehrlich during the first decade of this century. Their decline since then is due not so much to the failure of metal complexes as it is to the success of organic chemistry and biochemistry, marked by the overwhelming triumph of the sulphonamides in the 1930's followed shortly after by the discovery of the anti-bacterial antibiotics.

The bias towards organic materials is seen in the screening programme for new anti-cancer agents instituted with vigour in the mid 1950's at the National Cancer Institute in the USA. In the first 15 years of operation, of approximately 300,000 substances screened, about one half were characterised compounds. Of these only a handful were inorganic compounds and only a slightly larger number were metal organic complexes.

The discovery of the potent anti-tumour activity of platinum 4 years ago led to extensive screening of related compounds by the National Cancer Institute, the Chester Beatty Research Institute and other centres. As the results of many tests became available, a picture developed of a group of compounds with a wide spectrum of anti-tumour action and with potency as high as the best tumour chemotherapeutic agents.

This is just the beginning. With the application of quantum mechanics and ligand field theories, the chemical knowledge of metal co-ordination complexes has become quite sophisticated. At the same time there is a heightened awareness of the role of metals in normal biochemical functions. Taking all together an exciting new field is now emerging under the conglomerate title „bioinorganic chemistry".

Although there may be a reluctance to use heavy metals in man because of the widely held opinion that their soluble derivatives are general biological poisons, and although the kidney toxicity of *cis*-platinum (II) diamminedichloride has proved disappointing we know from experience that no general rules can be formulated for the biological properties of heavy metals. It is true that bare ions of heavy metals react readily and nonspecifically with nitrogen, sulphur, oxygen and other groups and can create havoc in the cell. However we are not concerned with bare ions here; we are concerned with metals locked in a tight ligand sheath, and in these, the ligand exchange reactions can be highly controlled and very specific.

A few hundred metal coordination complexes have now been tested in animals and they exhibit a rich diversity of effects and toxic levels that rival anything in the purely organic realm. For example, *cis*-platinum (II) diamminedichloride is an active anti-tumor agent in animals at concentrations ranging upwards from less than 1 mg/

kg. The LD_{50} is 13 mg/kg. The two chlorides can be replaced by a bidentate oxalato or malonato ligand and the complexes are as good or better than the first, and about an order of magnitude less toxic. Similarly, we can replace the two ammine groups by the bidentate ethylenediamine ligand and have as good or better activity than the first. Combine the two changes to make oxalatoethylenediamineplatinum (II) and suddenly we have an extremely toxic, fast acting neuromuscular poison with no anti-tumor activity even at tolerable dose levels. Now form malonatoethylenediamineplatinum (II), which has one extra carbon in the closed ring over the previous, and we again have a relatively mild, highly active anti-tumor agent. Therefore, small modifications in the molecular structure about the metal can produce radical changes in the biological effects.

Tobe, Connors and their co-workers have shown that the *cis*-dichloro-*bis*-cylohexylamineplatinum (II) congener has an LD_{50} of over 3,200 mg/kg, while the effective curative dose (ED_{90}) for the ADJ/PC6 tumor is of the same order of magnitude as for *cis*-platinum (II) diamminedichloride. This yields a therapeutic index well over 200, one of the highest ever reported. Some of the very soluble "platinum blues" are tolerated in mice at levels in excess of 800 mg/kg. Some preliminary work in our laboratory suggests that kidney toxicity may be negligible for these latter complexes. These, and numerous other examples, force the conclusion that the generalized fears concerning the use of heavy metals, in this context, are not valid.

cis-platinum (II) diamminedichloride has dominated the scene by virtue of being chosen the first to enter into clinical trials, but it is evident that other metal coordination complexes already exist, with preferred characteristics, as potential choices for second and third generation drugs. These two new subclasses have been discovered only within the last 2 years. This suggests that many more medically interesting complexes are available for the seeking, and recommends a deeper involvement of coordination chemists in synthesizing, and submitting for testing, the widest possible variety of complexes.

One of the delights of science is to be a committed worker in, and observer of, the evaluation of a new field of research. It is doubly delightful when, as in this field, the characteristic spirit is warm cooperation and a generous sharing among the scores of scientists and the industrial companies involved. I have a deep debt of gratitude to all of these latter groups and in particular to Engelhard Industries, Matthey Bishop Incorporated, and Rustenburg Platinum Mines, Ltd. for their continued generous support of this research, and in addition to Abbott Laboratories, Engelhard Industries, Ltd. and Kyowa Hakko Kogyo Co., Ltd. for support of this Second International Symposium on Platinum Coordination Complexes in Cancer Chemotherapy. Finally, a warm note of thanks to Ms. Ritta Rosenberg, who so beautifully organized the Symposium, and to Dr. R. J. P. Williams, our gracious host at Oxford University.

Metal-Binding to Bio-polymers: Some Thoughts on Anti-tumour Activity

R. J. P. Williams

Introduction

The effect of a large series of platinum complexes upon growing tumour cells has now been examined. As a consequence the active anti-tumour compounds are well-separated from the inactive compounds. Unfortunately this knowledge has not yet been supported by an equivalent study either of the chemistry of platinum/biopolymer reactions or of their molecular biology. Consequently, and despite the fact that the discovery of these chemicals leads to the possibility of following anti-tumour activity with inorganic probes, we are not able to make specific statements about the functional significance of platinum at the molecular level. The worst feature of all is the decreasing effort at the micro-biological level. I return to this point later.

Firstly let me say what has been discovered by examining the binding of many metal ions to biopolymers and thence what may be especially significant about the platinum compounds.

Metal-Binding to Proteins, RNA and DNA

All metals bind to proteins at high enough concentration (dose). We shall be interested only in strong binding, i. e. binding constants $\geqq 10^6$. This restriction eliminates very few metals from our discussion — only sodium, potassium and perhaps magnesium. The next step is to divide the metals of very different chemistry from one another. We shall choose to divide as follows:

a) Ca(II), UO_2(II), Ln(III), i. e. Group II A and III A of the Periodic Table and comparable cations.

b) (i) Lighter transition-metal ions such as Fe, Cu, (ii) Heavier transition-metal ions such as Pt, Ir.

c) The B-subgroup metal ions of the Periodic Table such as Zn, Pb, Hg.

Group II A and III A Metal Ions

The binding of the first division of cations is relatively easily described. The known binding sites of these metal ions taken from protein crystal structures are given in Table 1. It is invariably the case that the metals seek out multicarboxylate

Table 1. Binding sites for calcium and lanthanides

Protein	Site	Binding characteristics [a]
Carp albumin	Four carboxylate residues (Ca)	*s, f*
Thermolysin	Three carboxylate residues (Ca)	*m, f*
	There are two sites for Ca(II) and one for Ln(III)	
Concanavilin	Two carboxylate residues (Ca)	*m, f*
Bacterial nuclease	Three carboxylate residues (Ca or Ln)	*s, f*
Insulin	Carboxylates of different chains	*w, f*
Flavodoxin	Two (?) carboxylates	*w, f*
Lysozyme	Two carboxylates	*w, f*

Notes. 1. There is a wealth of data on UO_2(II) binding showing that it too binds mainly to carboxylate groups. 2. Pb(II) falls between Group II A and Group B elements and sometimes binds to carboxylate sites. Compare the close resemblance of K and Tl(I) binding sites in subtilisin and in the chymotrypsin inhibitor.

[a] strong, *s*; medium, *m*; and weak, *w*; binding and *f*; fast exchange.

centres. There is selectivity of binding even within the division but this is based on ion size and *not* on covalence. We shall not describe these effects further at it is clear enough that platinum does not compete at these sites and that these metal ions do not compete for the prefered platinum binding sites (see below).

The A-subgroup metal ions bind to RNA and DNA at the phosphate groups. They are undoubtedly used as required cross-linking agents in tRNA and rRNA and perhaps in DNA. However their binding is weak and falls outside our discussion. Once again there is little or no reason to suppose that this binding is related to or affected by platinum binding (or vice versa) in bio-polymers, see later.

Lighter Transition-Metal Binding Sites

The sites of binding of these metal ions *in vivo* are known in a very large number of proteins, see Table 2. No general remarks about the chemical character of the sites can be made for it is clear that the metal ions can bind *very selectively*. The sites are composed of a variety of binding groups of proteins e.g. Conalbumin — phenolate oxygen and histidine nitrogen; Ferredoxin — cysteine and sulphide sulphur; Myoglobin — histidine nitrogen.

The geometries as well as the binding groups are of a large variety of types. One common feature is outstandingly important: the metal ions are bound more or less permanently and strongly only if they are bound to at least *three* protein or coenzyme groups. For first row transition metal ions such as iron, copper and cobalt it is only through the formation of several bonds that a strong enough binding centre to retain them can be built, given the low concentration of these metal ions available to biological systems. Undoubtedly platinum could compete for such centres, mercury(II) binding to them has been shown frequently, but as *exchange* is very slow indeed this possibility can be ruled out. Platinum does not act by blocking metallo-enzymes.

Table 2. Binding sites for lighter transition metals

Protein	Site	Binding characteristics [a]
Conalbumin	Nitrogen and three tyrosines (Fe)	*s*, *s*
Concanavilin	Histidine and carboxylate (Mn)	*m*, *f*
Ferredoxin	Cysteine (two) (Fe) and sulphides	*s*, *s*
Rubredoxin	Cysteine (four) (Fe)	*s*, *s*
B_{12} enzymes	5 nitrogens of coenzyme and one carbon (Co)	*s*, *s*
Heme proteins	4 nitrogens of coenzyme and variously nitrogen or sulphur protein ligands (Fe)	*s*, *s*

[a] Strong binding, *s*; medium binding, *m*; and fast exchange, *f*; slow exchange, *s*. Binding comes first.

Finally we note that these transition metals are not associated with DNA or RNA *in vivo* to any discernable degree. We will discuss the heavier transition metals later.

Lighter transition metal ions will bind *in vitro* to a large number of additional biopolymer centres. This binding is unlikely to be of much importance *in vivo* as it is usually relatively weak and to surface groups of proteins, from which the ions can exchange rapidly. The binding would be prevented by the binding of the metal ions to a large number of small molecules which effectively buffer the biological system. For example the binding of copper(II) to amino-acids will reduce very considerably the free Cu(II) concentration. We note however that should the carrier molecules and proteins for iron and copper become saturated then the effects of the free transition metal ions can be extremely deleterious. These metals act as redox catalysts as well as directly as binding agents and have a very high toxicity. Relatively speaking their binding to RNA and DNA is much less noticable than to proteins. We note that both RNA and DNA can provide two binding centres easily, but not three in contrast with the specific metal ion sites found in metallo-enzymes and metalloproteins. An outstanding result is then that proteins provide better binding groups due to their more flexible geometry than does DNA or RNA. Single stranded DNA or RNA supplies no strong binding sites.

B-Subgroup Metal Ions

The binding of zinc, Table 3, to proteins is well studied. The sites, as for light transition metal ions, are nitrogen, carboxylate and sulphur. The geometric disposition of the groups is important. Three such groups are required to hold the zinc in a protein strongly and to reduce exchange. The situation *in vitro* is very similar to that for the light transition metals. Zinc does not occur in RNA or DNA.

Mercury(II) is very different. It does not occur naturally in biopolymers of course but it binds very strongly indeed even to single sulphydryl groups. Exchange is then relatively slow. Mercurials are favourite heavy atoms for X-ray crystallography. Here we see a major distinction between heavy and light B-subgroup metal ions — extreme strength of binding to one or two sulphur (and perhaps nitrogen or even carbon) groups can make a more or less permanent binding site for *heavy* metal ions. Exchange (on the second or minute scale) is slow. The importance of these observations becomes obvious below.

The heavy B-metals also bind to RNA and DNA. Elsewhere we have discussed this binding [5]. It involves one or two nitrogen bases, is of high binding strength, and exchange is not necessarily fast.

We can enquire "what is the difference between zinc and mercury?" The most important distinction lies in the bonding. At a very naive level of interpretation the mercury binding (and this applies to all heavy transition metal ions such as platinum)

Table 3. Binding sites for zinc

Protein	Site	Binding characteristics [a]
Carboxypeptidase	2 histidines, one carboxylate	*s*, *s*
Insulin	2 histidines (on each chain)	*m*, *f*
Carbonic anhydrase	3 histidines	*s*, *s*
Alcohol dehydrogenase	1 cysteine plus other groups	*s*, *s*

[a] Binding (first) is shown as *s* strong or *m* medium and exchange rate is shown as *s* slow or *f* fast.

is more covalent. A second feature is that the binding to mercury is stereochemically very limited. Mercury forms two linear bonds. These two factors acting together cause mercury to be a very effective blocking agent of enzyme systems for mercury is largely restricted to $-SH$ group attack. The effect of other heavy transition-metals such as Tl, Pb, As and Sb would appear to be similar in that they bind in a more covalent manner but the geometric factors are rather different. These metals too are very effective poisons (and drugs), but they bind to a much wider variety of centres.

Heavy Transition Metal Ions (Pt)

The binding of these metals, especially platinum, is now very well studied through the work of X-ray crystallographers. The compounds used $PtCl_4$, $AuCl_4$, are usually bound on the outsides of proteins. They are of course charged and they are very bulky. The main centres attacked, Table 4, are sulphur groups but nitrogen bases are also involved. The binding is through one or two protein ligands.

In proteins binding of Pt(II) to a single centre [compare mercury(II)] is sufficient to make a valuable derivative for X-ray structure determination purposes. Methionine, cysteine, cystine are possible groups. The binding is not easily reversible e. g. in ribonuclease. At higher concentrations platinum compounds bind to very many protein sites but this binding depends too on the exact nature of the platinum compound. For a variety of reasons which are discussed by others in this book, it is my opinion that although the anti-tumour agent *cis*-$[Pt(II)Cl_2(NH_3)_2]$ will attack a great variety of proteins this is not the attack which generates anti-tumour activity.

An obvious feature of platinum binding is its covalence and this it shares with mercury. However there are two marked differences between the two metal ions. Firstly there is the stereochemistry. Hg(II) forms two linear bonds while Pt(II) forms four bonds in a square. Secondly Pt(II) exchanges nitrogen groups slowly so that effectively sites on the Pt(II) can be made inert. *Cis*-[Pt(II)$Cl_2(NH_3)_2$] is a very different reagent from *trans*-[Pt(II)$Cl_2(NH_3)_2$]. Unlike the lighter transition metals, there is *very little* flexibility in the geometry around the metal, and in order to obtain a bound slow-exchange situation to a bio-polymer *one or two* binding sites (not *three*) are sufficient. It is not surprising that strong binding of a compound such as *cis*-[Pt(II)$Cl_2(NH_3)_2$] is very specific in biological systems for it will require a specific chelate geometry, as well as binding weakly and more generally through one group.

Table 4. $[PtCl_4]^{2-}$ Binding sites in proteins

Protein	Site
Chymotrypsin	—S—S—, Met, NH_2 terminal
Cytochrome-c	Met, His
Ribonuclease	Met
Subtilisin	Met, His
Carboxypeptidase	—S—S—, Met, His, NH_2 terminal
Human-γ-globulin	same as Hg site
Concanavilin	same as Hg site
Prealbumin	same as Hg site
Thermolysin	His
Triose phosphate isomerase	same as Hg site —SH and other sites

Notes. 1. High concentration of amines can be used to remove Pt from a binding site in some proteins. 2. The site attacked depends upon the buffer medium in which the Pt is dissolved. Pt compounds bind to many simple groups such as phosphates. I am grateful to Prof. D. C. Phillips and Dr. G. Petsko for help in compiling this table. 3. There is some data to show that elements such as Pd, Ir, Au(III) bind in a similar way to Pt.

Table 5. The nature of coordination compounds in biology

Metal ion	Type of ligand as partner	Number of partners for strong binding	Exchange rate from site	Stereochem. demand
Na, K	Oxygen (O)	≥ 4	Fast	Nil
Mg, Ca	Oxygen anions	≥ 4	Medium	Nil
Zn	O, N, S	≥ 3	Slow	Little
Fe, Co	O, N, S	≥ 3	Slow	Medium
Hg	N, S	2.1	Slow	Strong
Pt	N, S	2.1	Slow	Strong
Carbonium ion	N, S, (O)	1	Slow	Strong

Note. The above table applies to exchange from a protein. It would be unwise to use it as more than a general guide but it does show how the chemical ideas currently used in the understanding of organic mechanisms (last line) are closely parallel to those needed for an understanding of platinum chemistry.

Through the work of a large number of research groups, reported later in the symposium, it seems likely that *cis*-[Pt(II)] is a cross-linking reagent for DNA. It is clear that if such cross-links are to be effective the Pt(II) compound formed must have kinetic stability. Reaction with basic nitrogen atoms in DNA would give such stability but this reaction also produces a product which has a particular structure. Such an attack would be difficult to mimic with other metal complexes which are not closely related to Pt(II) despite the fact that many metals will attack the basic nitrogen atoms in one way or another. We stress the reasons again, Table 5.

a) Pt(II) will bind strongly when bound through *two* nitrogen residues.

b) The binding is likely to be only *slowly* reversible.

c) Stereochemical selectivity is built in to the reagent as the two NH_3 groups of $[(NH_3)_2PtCl_2]$ will not exchange.

d) Pt(II) has common reactions with other compounds which attack DNA, Table 5.

Given that the site of attack is specific, how does the platinum find the site? What could this specific (critical) site be?

Entry of Pt(II) into Cells

Our knowledge of the activity of the Pt(II) compounds is also hindered by a lack of knowledge of a great deal of Pt(II) chemistry with biologically interesting groups. Before any attack on DNA can occur the *cis*-dichloro-compound of Pt(II) is likely to undergo multiple rapid substitutions. It could well be that in the high chloride medium of the blood stream the reactions are only the exchange of one chloride for another. In cells the chloride ion concentration is much lower and under such conditions chloride will be replaced by such groups as phosphate anions. Again, passage across the cell membrane may be very restricted for an ion such as $[Pt(H_2O)_2(NH_3)_2]^{2+}$, (while $[PtCl_2(NH_3)_2]$ and such molecules as $[PtCl_2(NH_2R)_2]$ may penetrate readily), but inside the cell it could well be that the dichloride is not the strongest electrophile. We can make few comments about these points as yet. I believe that Pt-phosphate bonding inside cells also needs to be considered. Its exchange would be moderately fast.

Why Platinum?

The question as to why it is platinum that is active, and platinum in a particular form, is now answerable using the summary table, Table 5. A set of chemical parameters is optimised somewhere close to platinum in the periodic table but complexes of elements such as Au, Hg, Ir, Rh, Os, Ru, Re in particular could form compounds which might mimic *cis*-$[Pt(NH_3)_2Cl_2]$ quite well, although it could well be difficult to find a very close relative of this particular metal complex using another metal.

Now while we may suppose that it is the cross-linking function of the bi-functional platinum compounds which leads to anti-tumour activity (see later discussion in this symposium) it would be a mistake to classify the platinum drugs together with other bi-functional reagents such as the di-alkylating agents also used to cross-link DNA. Undoubtedly there are many similarities in the general pattern of activity of the two series of drugs but the differences in chemistry and stereochemistry are very large indeed. In fact I do not regard cross-linking as a readily definable term. In order to bring out this point and to establish an attitude of mind to "cross-linking" I must first give a very brief comparative account of the way in which I see platinum drugs in relationship to other drugs used in cancer chemotherapy.

A Comparative Approach to Platinum Activity

In this final section I shall try to stimulate some discussion as to why the action of these platinum compounds is so biochemically specific. I want to stress the importance of microbiological studies. I shall start from the general premise that induction of provirus (lysogeny) and anti-tumour activity have much in common, but

Table 6. Groups which attack DNA

Alkylating agents	Epoxides, alkyl halides (note also Pt), Nitrosoalkylureas, sulphur and nitrogen mustard.
Intercalating agents	Polycyclic hydrocarbons, sterols, Mitomycin c, nalidixic acid.
Radical agents	$OH^{\cdot}$ etc., from NH_2OH, H_2O_2, etc., $h\nu$, X-rays.
H-bond agents	Peptides such as Distamycin.
Biological incorporation	Bromouracil etc.

that agents which are effective to some extent in both activities have very variable effects in (i) stimulating elongated (filamentous) growth of bacterial cells, (ii) causing mutation. In the Tables 6, 7 and 8, I have attempted to make some comparative remarks about different chemicals which have these effects. From the Tables it is clear that filamentous growth is only to be associated with anti-tumour activity in a very vague way and there must be different ways of causing filamentous growth both related to and dissociated from any effect on DNA. (It was merely fortunate that Rosenberg discovered that the platinum compounds act as anti-tumour agents because he observed filamentous growth in *E. Coli.*) Again mutagenic agents must act at the level of DNA, yet clearly effective mutagenic agents are not good anti-tumour agents nor do they cause pro-virus induction very effectively. Loveless [3] pointed to the fact that the most powerful inducing agents were usually the poorest mutagens and *vice versa,* although nearly all agents which had one effect also showed the other to some degree, Table 9.

As I see it, the platinum compounds fall in the group of good inducers and poor mutagens, although the last part of the statement is based upon the absence of evidence for mutagenic effects. This division of the compounds corresponds closely with the distinction between two types of agent which attack DNA — those that "cross-link" and those that do not, Table 10. I conclude from the nature of the chemicals that good "cross-linking" agents are poor mutagens but good inducers, whereas good mutagens attack single bases and do not act as inducers. I believe then that platinum is again in the first group. To improve the general line of this discussion I have

Table 7. Properties of compounds which interact with DNA [a]

Agent	Anti-tumour	Mutagenic	Carcino-genic	Prophage produced	Fila-ments	Synthesis of colicin*
UV	?	++	+	+	+	+
X-rays	+	+	+	+		−
Di-(2-chloroethyl)-methylamine	+	+	+	+	+	+
2-chloroethyl-dimethylamine	(0)	0	0	0	−	−
H_2O_2 in inorg. media		0	−	0		−
H_2O_2 in org. media		(+)	−	+	?	+
t-Bu-O_2H		(+)	+	+		+
2,4,6-*tris*-ethyleneimino-1,3,5-triazine		+	+	+		−
Butadiene-1:3-diepoxide		+	+	+		−
Ethylurethan Water soluble		+	+	0		−
1:4-dimethanesulphonyloxy-butane		−	+	0		−
9:10-dimethylbenzanthracene	0	+	+	0		−
2-acetaminofluorene	0	+	+	0		− Not
Methylcholanthrene	0	+	+	0		− water
β-naphthylamine	0	+	+	0		− soluble
benzidine	0	+	+	0		−
$Pt(NH_3)_2Cl_2$	+	? −	? −	+	+	?

[a] From Lwoff [4].
* Highly specific proteins which kill other bacilli.

drawn up in Table 10 lists of reagents indicating the *strength of their cross-linking* ability. It is a simple matter to compare this list with mutagenic action or with lysogeny. Could it be that the effective anti-tumour agents are just those which arrest DNA synthesis most effectively? This could be brought about either through direct attack on DNA (Pt), or through attack on the cell membrane for a cell is an intercommunicating system. For my part I feel that the biochemistry of the above compounds is too selective to represent an attack on chemical groups in the membrane.

When comparing the agents which produce cancer with those that "cure" it I have often heard quoted a statement called Haddow's Paradox — "The agents which cause cancer also cure it" [2]. Whether Haddow ever said precisely this or not I have been unable to ascertain but it should not be generally true. It seems to me that cancer producing and anti-cancer agents should differ in much the way that mutagens differ from phage-inducing agents. The difference will rest in the *strength of their cross-linking*, i. e. there should be a difference of degree if not of kind.

There should then be a wide spectrum of reagents which will be effective in both attacks, but the extremes of the spectrum will be represented by anti-tumour agents

Table 8

1. Inducing agents not mutagenic or carcinogenic and having no effect as anti-tumour agents (no attack on DNA)
   a) Certain non-metabolised sugars
   b) Pressure
   c) Colicins
2. Mutagenic agents not known to cause induction of lysogenic bacteria (all weak monofunctional agents)
   a) Monofunctional alkylating agents
   b) Aromatic hydrocarbons
   c) Aminouracil (compare bromouracil)

Table 9. Classification of reagents which are thought to attack DNA

Compounds with a high mutagenic rate relative to lysogenic lysis or bacteriocidal rate	Compounds with a low mutagenic rate relative to lysogenic lysis or bacteriocidal rate
1. Mono-alkylating agents	1. Multi-functional alkylating agents
2. Amino-substituted bases	2. Bromo-substituted bases
3. Low energy radiation	3. High energy radiation
4. Simple monofunctional intercalation agents — ethidium bromide polycyclic hydrocarbons	4. Complex multifunctional intercalation agents — Mitomycin C
5. $MnCl_2$	5. *cis*-$Pt(NH_3)_2Cl_2$
	6. Radical producers (proflavine, H_2O_2) compare hydroxylamine with hydrazine
	7. Polypeptides, Distamycin A

Table 10. Strength of cross-linking

Zero	Poor	Medium	Strong
RCH_2Br	Aromatic hydrocarbons	Azaserine	$(.CH_2Br)_2$
Monoepoxides	Mitomycin C	Antibiotic peptides	Di-epoxides
	Mn(II)		$Pt(NH_3)_2Cl_2$
	Weak X-ray ($h\nu$)		Strong X-ray ($h\nu$)

incapable of causing mutations or cancer (and which stop DNA synthesis entirely), while the carcinogens (usually mutagens) should alter DNA but in no sense stop DNA synthesis. The paradox comes about as most agents which attack DNA are *weak* cross-linking agents and so give a mixture of two responses. It is unfortunately the case that, if my conclusion is correct, the best cross-linking anti-tumour agent is then an outright killer of a cell. It may well be thought better then to use anti-tumour agents which recognise specifically the outside of the tumour cell and so stop DNA synthesis in those cells only, rather than using a reagent (such as Pt) which must attack all cells' DNA, but with the hope that the DNA of a growing tumour cell is more open to attack than the DNA of the normal cell. I do not accept this point. I believe that we should culture tumour cells and normal cells and under conditions of contact inhibition of the normal cells see which type of cell is more readily labelled by radio-active platinum from a variety of compounds. Could it be that the entry of platinum into cells can be controlled so that the platinum compounds can be made to recognise tumour cells? The kinetics of replacement of different groups around Pt(II) are under the control of the other ligands present in a very subtle way.

A Note on Cross-Linking

Any chemical which interacts *across* the strands of DNA is a cross-linking agent. The nature of the chemical forces which are involved can vary from VAN DER WAALS, through hydrogen-bonding and ionic links, to covalent bonds. We can place such bonding in a rough order of kinetic stability which is VAN DER WAALS $<$ hydrogen-bonding $<$ ionic links $<$ covalent bonding. This list can be related to a series of chemicals: aromatic hydrocarbons $<$ peptides $<$ di-alkylating agents, Table 10. In the list we can place metal ions remembering, Table 5, the changing covalent character of their bonds. In this sense Pt forms strong cross-links. Current methods of detecting cross-links will often only detect those of high kinetic stability. Is it true that only such cross-links are important in drug function or is there a graded series of effectiveness following the gradient of cross-linking strength? Given the different modes of possible attachment of different drugs I fail to see why different cross-linking chemicals should not be synergistic in action. I conclude by stating firmly that the type of cross-link formed by the platinum compounds is of a new kind and differs from di-alkylating agents much as both of them differ from antibiotic peptides. It is certainly very different from the *ionic* cross-linking between strands of DNA which is brought about by such cations as $[Mg(H_2O)_6]^{2+}$ and which is essential for replication of DNA. All degrees of cross-linking of strands are possible as there are many ways of cross-linking and each mode can be varied in strength.

In such a general lecture as this references are rather useless and I have therefore given five general starting points only in the reference list. I trust this will cause no offence for my role is educational.

Acknowledgments

I wish to thank Drs. DROBNIK, ROSENBERG and RESLOVA for many discussions over several years. My own work on Metal Ions in Biological Systems has the support of both The Medical Research Council and The Science Research Council of England. I have been greatly assisted by several X-ray Crystallographers who have given me access to data. Only by working back through the references in The Cold Spring Harbor Symposium [1] can the metal-binding sites be traced.

References

1. Cold Spring Harbor Symposium on Quantitative Biology. Structure and function of proteins at the three dimensional level. **36** (1972).
2. HADDOW, A.: Cellular inhibition and the origin of cancer. Acta Un. int. Cancr. **3**, 342 (1938).
3. LOVELESS, A.: Genetic and allied effects of alkylating agents. London: Butterworths 1966.
4. LWOFF, A.: Lysogeny. Bact. Rev. **17**, 269 (1953).
5. THOMSON, A. J., WILLIAMS, R. J. P., RESLOVA, S.: The chemistry of complexes related to *cis*-$Pt(NH_3)_2Cl_2$, an anti-tumour drug. Structure and Bonding **11**, 1746 (1972).

Chemistry of Co-ordination Complexes

M. J. CLEARE (Rapporteur)

1. Some kinetic studies on derivatives of potential anti-tumour Pt(II) complexes.
   P. BRADDOCK, A. KHOKHAR, R. ROMEO, M. L. TOBE, University College, London, England.
2. Studies of platinum interactions with nucleosides.
   P. C. KONG, T. THEOPHANIDES, Université de Montréal, Montreal/Canada.
3. Platinum Blues.
   I. A. G. ROOS, A. J. THOMSON, University of East Anglia, Norwich/England.
4. The molecular structure of μ-pyrophosphato-bis[diammino-platinum(II)]. A novel pyrophosphate bridged dimer involving probable Pt-Pt metal bonding.
   J. A. STANKO, Pennsylvania State University, Pennsylvania/USA.
5. The crystal structure of dichloro(ethylenediamine)platinum(II), [$PtenCl_2$].
   J. IBALL, S. N. SCRIMGEOUR, University of Dundee, Dundee, Scotland.
6. Symmetrical and unsymmetrical addition to platinum compounds.
   J. V. RUND, University of Arizona, Arizona, USA.
7. The chemistry and anti-tumour, anti-viral, anti-bacterial and anti-fungal screening of 15 palladium and 2 nickel complexes.
   R. D. GRAHAM, D. R. WILLIAMS, University of St. Andrews, St. Andrews, Scotland.
8. Inhibition of growth and cell division in *Escherichia coli* by compounds containing rhodium and some comments on the related effects of compounds containing platinum.
   R. D. GILLARD, K. HARRISON, I. M. MATHER, University of Kent, Canterbury/Kent, England.
9. Studies on the anti-tumour activity of group VIII transition metal complexes Part II Rhodium and Iridium amine complexes.
   M. J. CLEARE, T. A. CONNORS, J. D. HOESCHELE, B. ROSENBERG, L. VAN CAMP, Johnson Matthey and Co. Ltd., Wembley, Middlesex, England
10. The synthesis of metal chelates as potential anti-cancer substances.
    A. J. CHARLSON, Macquarie University, New South Wales, Australia.

Introduction

The emphasis which should be put on material for the Co-ordination Chemistry Section is open to individual interpretation, and for this presentation the major theme is the relationship between the chemical structure and properties of the compounds and their anti-tumour activities (i. e. Structure—Activity Relationships). There is some overlap with Session II and several papers could have been presented in either section. The intention is to summarise briefly the known relationships and existing data, and then to discuss research which is at present in progress as described in the submitted papers. Where possible an attempt is made to indicate areas which might usefully be pursued for gaining a better understanding of this new and fascinating area of inorganic research. It should be pointed out that this Session also contains papers which discuss transition metals other than platinum.

One of the major difficulties in collating animal testing data has been the non-comparability of test systems (i. e. strain of animals, type of tumour and its origin). This is partially due to the nature of the disease and also to the imperfections of the transplantable test tumour system. Indeed, the activity of a compound should be considered as specific to the defined test system until proven otherwise. However, activity against any test system allows one to consider the compound as a potential anti-tumour agent, whether or not that potential is justified later by further testing. As far as possible comparisons will be limited to results within a common system, at the same time indicating if broader testing studies have been attempted.

I. Platinum(II) Complexes

Anti-tumour activity appears to be confined to neutral complexes with a *cis* configuration, or more relevantly, *cis* leaving ligands. A comparison of *cis* and *trans* compounds of the type $[PtA_2X_2]$ shows activity either a) only for *cis* and not *trans* or b) not in either *cis* or *trans*. No case of activity in *trans* but not in *cis* has been found (Table 1).

In general charged complexes show no activity even when they have *cis* leaving groups (particularly chlorides). It has been established that the nature (leaving ability) of the X ligands is important. Two main classes have been found to be successful:

a) Monodentate anionic ligands of intermediate leaving ability — mainly Cl^- but also Br^- (Table 2).

b) Bidentate carboxylate ligands — oxalate, malonates (Table 3).

Those in b) are not expected to be good leaving groups so some active mechanism of removal may be operating. In many cases an anti-tumour active chloroamine complex will have an active malonate analogue [1].

Although the amine (A) ligands are expected to have a secondary effect on the reactivity of the complexes (excepting cases involving extreme steric hindrance), they have been found to play a vital role in the anti-tumour property. Against the S. 180 tumour (which is less sensitive to variations in the Pt complexes) substitution of the H atoms in ammonia generally reduces the activity, but variations have not been reliably observed (Table 4). However, the more sensitive ADJ/PC6 tumour of the Chester Beatty Institute has shown some fascinating trends which are rather difficult to interpret in chemical terms [2]. Lower alkyl substitution appears to diminish selectivity against the tumour (as measured by the therapeutic index, T.I.) while heterocyclic and alicyclic amines show a great increase in this index, largely (but not entirely) due to greatly lowered toxicities. Most active compounds have an $I.D._{90}$

Table 1. Comparison of activities for *cis* and *trans* isomers

Sarcoma 180 [1] Complex		Solvent	Dose range mg/kg	Dose response	Toxic level mg/kg	T/C	Dose mg/kg
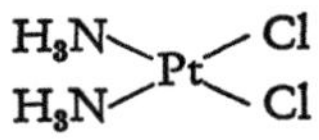	*cis*	S	0.5— 20	+	9	1	8
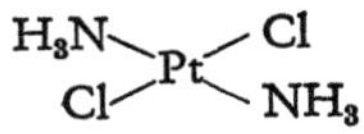	*trans*	S	2.5— 40 [b]	—	>40	85	2.5—40
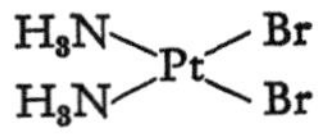	*cis*	B	5 — 20	+	15	30	14
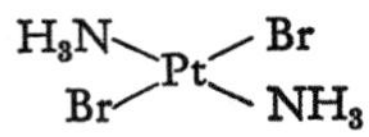	*trans*	B	10 — 40	—	>40	110	10 —40
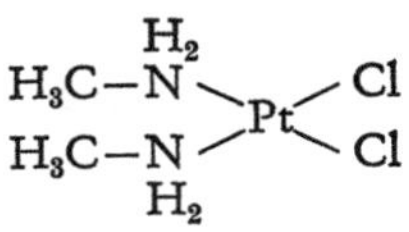	*cis*	S	10 — 30	+	12—20 [c]	25	15 [a]
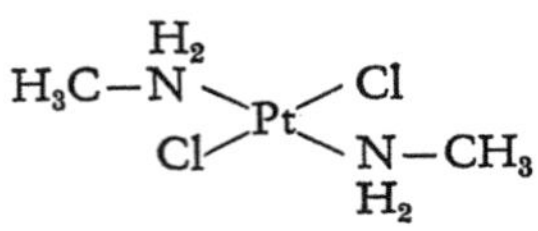	*trans*	S	5 —100	—	25	100	5 —20
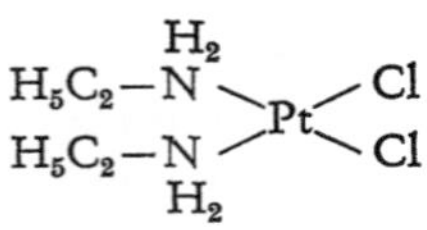	*cis*	S	5 — 50 [b]	+	45	14	40
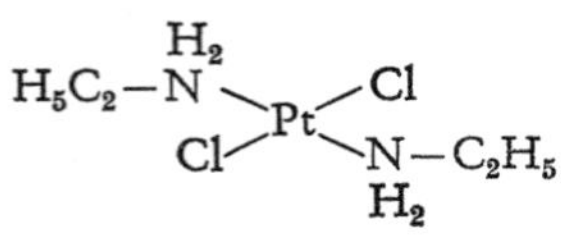	*trans*	SS	5 — 20	—	>20	105	5 —20

(Table 5) between 1—20 (often 1—10) mg/kg, while the L.D.$_{50}$ may vary from 13 (*cis*-Pt(II)) to several thousand mg/kg, although these figures could be distorted by very low aqueous solubilities.

Table 6 shows some recent tests on the ADJ/PC6 tumour [3]. The points of interest are:

a) When the alicyclic ring is removed one carbon away from the nitrogen then the T.I. is greatly reduced.

b) The presence of heterocyclic or alicyclic rings as part of a chelate (with their nitrogen atoms attached to Pt) also confers considerable activity and selectivity.

c) Branched alkyl primary amines give good selectivity with the *iso*amylamine complex outstanding.

The ADJ/PC6 tumour appears to be excellent for detecting subtle variations in activity with changing structures. The problem is to correlate these variations.

Professor M. L. TOBE and his co-workers have carried out kinetic studies on their active complexes to see if a relationship exists between the lability of the various ligands and the anti-tumour properties. Their paper (*Paper 1* — P. BRADDOCK, A.

Table 1 (continued)

ADJ/PC6 plasma cell tumour [2] Complex		Solvent	Dose range mg/kg	Dose response	LD_{50}	ID_{90}	T. I.
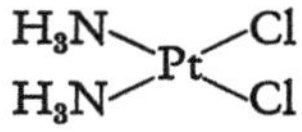	*cis*	A	0.1— 40	+	13.0	1.6	8.1
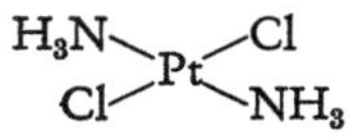	*trans*	A		—	27.0	>27.0	<1.0
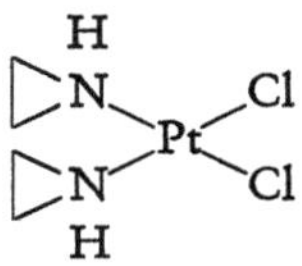	*cis*	A	2.5—160	+	56.5	2.6	21.7
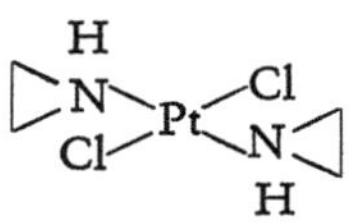	*trans*	A		—	18.0	>18.0	<1.0
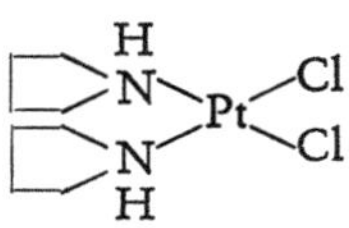	*cis*	A	6 —800	+	240	17.5	13.7
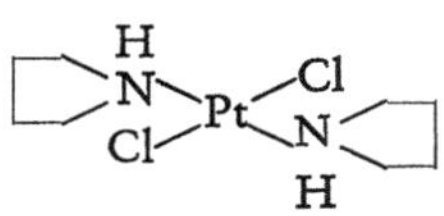	*trans*	A		—	72	>72	<1.0

[a] Only 66% survivors. [b] Slurry at higher concentrations. [c] Sporadic toxicity over this range.
A = Arachis oil; B = sodium bromide; DMSO = dimethyl sulphoxide; S = saline; SS = saline slurry; W = water; WS = water slurry.
This explanation of sign applies to tables 1—5 of this contribution.

Table 2. Variation of X in *cis*-$[Pt(NH_3)_2X_2]$ with sarcoma 180 tumour [1]

X	Solvent	Dose range mg/kg	Dose response	Toxic level mg/kg	T/C	Dose mg/kg
NO_3^-	W	6 — 12	—	7 [e]	54	6
NO_3^-	S	2.5— 12	+	11	8	10
H_2O [f]	W	2 — 20	—	5 [e]	—	—
Cl^-	S	0.5— 20	+	9	1	8
Br^-	B	5 — 20	+	15	30	14
Br^-	S	2 — 6 [d]	+	5—6 [d]	13	5
I^-	WS	10 — 25	—	> 25	110	10— 25
SCN^-	S	5 —100 [b]	—	~ 50	70	20— 35
NO_2^-	SS	5 —100	—	>100	99	5—100

[b] Slurry at higher concentrations. [d] Daily injections for 9 days. [e] Highly toxic — convulsions. [f] Cationic complex ion (2+).

Table 3. Activity of ammine complexes with chelated dicarboxylate ligands X in $[Pt(NH_3)_2X]$

Sarcoma 180 [1] X	Solvent	Dose range mg/kg	Dose response	Toxic level mg/kg	T/C	Dose mg/kg
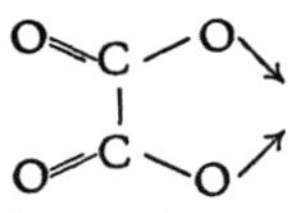 oxalate	DMSO (slurry)	5 — 20	+	16—20	9	15
	WS	12 — 18	+	17	24	14—16
	W	10 — 20 [g]	+	10	0	10 [a]
	W	0.5— 6 [d]	+	3— 4	25	2.5
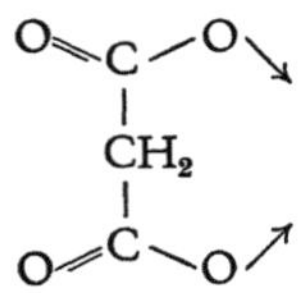 malonate	WS	10 — 60	+	35	7	30
	W	5 — 24 [g]	+	24	28	20—24
	W	1 — 7 [d]	+	8	12	7
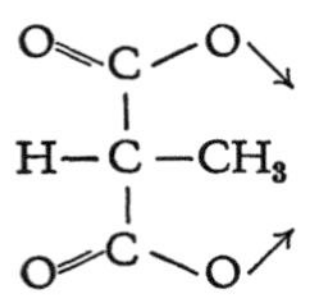 methyl malonate	W	10 — 80	+	65	7	60
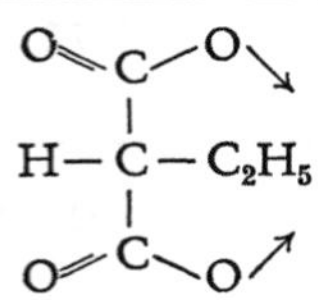 ethylmalonate	W	30 — 80	+	>80	17	70—80
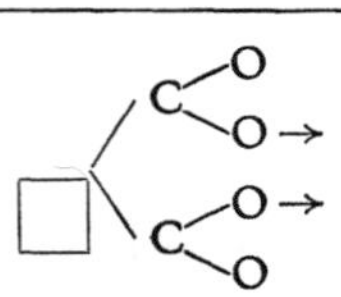 1,1 cyclobutane dicarboxylate	W	20 — 160	+	150	18	120

Khokhar, R. Romeo, M. L. Tobe) reports on studies on the alicyclic series *cis*-$[Pta_2Cl_2]$ where *a* is a primary amine with an alicyclic organic substituent, $C_nH_{2n-1}NH_2$ with n = 3 to 8. As we have seen, all these compounds are active with the maximum value for T.I. in the compounds with five and six membered rings (n=5—6; although the cyclohexylamine has the highest T.I., results from other tumour systems indicate that this high activity may be unique to the ADJ/PC6 tumour, whereas the cyclopentylamine appears to have a broader range of activity). Although it is generally agreed that the platinum complexes are likely to resemble alkylating agents and react via their chloro(anionic) groups, the amine ligand has a primary effect on the activity and these researchers are studying the following kinetic factors:

Table 3 (continued)

ADJ/PC6 plasma cell tumour [18] X	Solvent	Dose range mg/kg	Dose response	LD_{50}	ID_{90}	T. I.
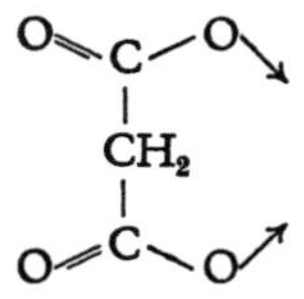 malonate	A	1 — 320	+	225	18.5	12.2
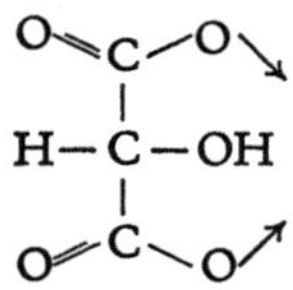 hydroxymalonate (tartronate)	A	2 —1500	+	150	4.9	30.6
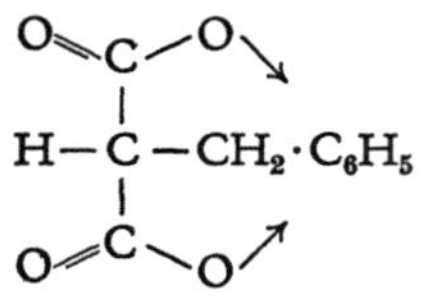 benzylmalonate	A	1 —1500	+	150	1.85	81.1
1,1 cyclobutane dicarboxylate	A	4 — 256	+	180	14.5	12.4

[a] 66% survivors only. [d] Daily injections for 9 days. [g] Multiple injections due to low solubility.

a) The way in which the amine affects the replacement of the Cl ligand (this is expected to be a small effect).

b) The way in which the nature of the amine affects its own lability.

c) *Cis* effects of one amine on another.

The replacement of one Cl group by DMSO introduces a strong *trans* labiliser and replacement of the amine *trans* to it is observed under mild conditions. Thus it is possible that replacement of Cl within the body by a similarly strong *trans* labilising ligand (e. g. a sulphur-containing protein) could labilise an amine and provide an alternative mechanistic path to that of direct reaction via the chloride ligands.

The reaction:

$$cis\text{-}[Pta_2(DMSO)Cl]^+ + Cl^- \rightleftharpoons cis\text{-}[Pta(DMSO)Cl_2] + a$$

lies well to the left and little reaction is observed until acid is added to protonate the released amine and prevent the reverse reaction (which seems to be fast). The rate is *first order* with respect to chloride and the nucleophile independent pathway (k_1) is generally absent or very small (except NH_3, n = 5 and n = 3, Table 7). This is un-

Table 4. Changes in activity on varying A in *cis*-$[PtA_2Cl_2]$

Sarcoma 180 [1] A	Solvent	Dose range mg/kg	Dose response	Toxic level mg/kg	T/C	Dose mg/kg
H_3N	S	4 — 15	+	9	3	8
CH_3NH_2	S	10 — 30	+	12—20 [e]	14	14 [a]
$(CH_3)_2NH$	S	30 — 150	+	~ 100	25	80
$C_2H_5NH_2$	S	5 — 50	+	~ 40	14	40
$(C_2H_5)_2NH$	SS	15 — 60	—	> 60	75	60
$HOC_2H_4NH_2$	S	20 — 225	+	~ 125	22	125
i-$C_3H_7NH_2$	SS	20 — 50	±	> 50	33	30
	WS	5 — 40	—	> 40	94	5
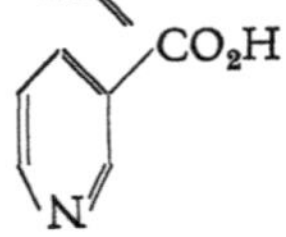	W [h]	4 — 80	±	> 80	51	40—60
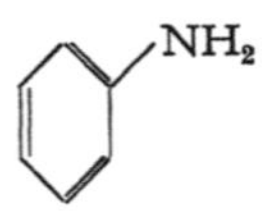	SS	10 — 50	+	> 50	~33	10—20

explained as pH stat and spectrophotometric measurements give very similar values for the rate constants, indicating that the amine release accompanies chloride uptake and does not precede it. Apart from NH_3, which is more inert than expected, there is a reasonably good free energy relationship between $\log k_2$ and pk_a of the conjugate acid (Fig. 1). The large slope indicates the importance of bond breaking in the transition state.

Fig. 2 compares potency ($I.D._{90}$) and toxicity ($L.D._{50}$) of the chloro complexes against the amine lability of the corresponding DMSO complex. The main conclusion to be drawn is that there is no relationship between the lability of the amine ligands and the anti-tumour property. Thus although this type of reaction is likely to occur, it dose not appear to make a significant contribution to the activity. In the cyclohexylamine case both *cis*-$[Pta_2(DMSO)Cl]Cl$ and $[Pt(a)Cl_2(DMSO)]$ are inactive, and are more toxic than the parent amine against the ADJ/PC6 screen.

Non-N Donors

Similar platinum(II) compounds with non-N donor ligands provide an interesting area for study. However, relatively little work has been reported and no papers on this subject were received except that *Paper 9* (A. J. Charlson) describes an inactive bis(thiosemicarbazone) complex.

Table 5

ADJ/PC 6 plasma cell tumour [2] A	Solvent	Dose range mg/kg	Dose response	LD_{50}	ID_{90}	T. I.
NH_3	A	0.1— 40	+	13.0	1.6	8.1
CH_3NH_2	A		−	18.5	18.5	1.0
$ClC_2H_4NH_2$	A		+	45.0	17.5	2.6
	A	2.5— 160	+	56.5	2.6	21.7
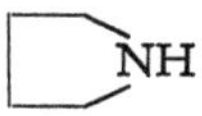	A	3 — 200	+	141	10.8	13.1
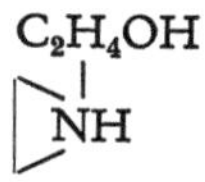	A		−	90	>90	< 1.0
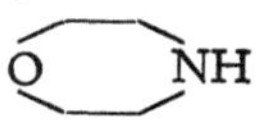	A		−	18	>18	< 1.0
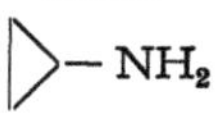	A	1 — 80	+	56.5	2.3	24.6
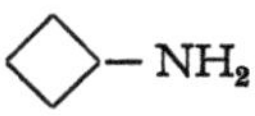	A	6 — 750	+	90	2.9	31.0
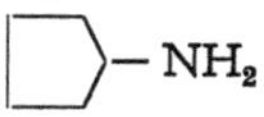	A	1 —3200	+	565.6	2.4	235.7
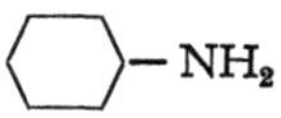	A	1 —3200	+	>3200	12	>267
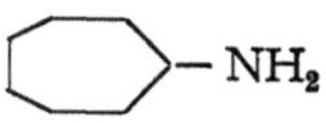	A	5 — 625	+	> 625	18	> 35

[a] 66% survivors. [c] Variable toxicity. [h] Injected as sodium salt.

II. Platinum Metabolite Complexes

Only a brief discussion of this area is possible as it overlaps considerably with Session II. Studies on the chemistry of such complexes are obviously desirable in terms of possible *in vivo* reactions and mechanisms of anti-tumour action. Here we are considering a few compounds which have been isolated and characterised.

Paper 2 (P. C. KONG, T. THEOPHANIDES) describes the reactions of some anti-tumour compounds, such as *cis*-$[Pt(NH_3)_2Cl_2]$ and $[Pt(en)Cl_2]$, with DNA nucleosides. Analysed complexes of the type $[Pta_2(nucl)_2]Cl_2$ have been obtained with guanosine and inosine. NMR spectra on $[Pten(nucl)_2]^{2+}$ (nucl = guanosine and inosine) indicate N7 binding which is in agreement with other studies. Fig. 3 shows the NMR spectrum of guanosine in DMSO and the complex in D_2O with the assignments. It is assumed that the purine protons which are shifted downfield significantly are nearer to the co-ordinating position (i. e. H_8); the spectra of inosine and the inosine complex show similar shifts (Fig. 4). In each case H_8 is weakly coupled to platinum. The spectrum of the inosine complex changed with time in DMSO which is not surprising as DMSO is an S-bonding ligand (see Paper 1).

Table 6. Recent ADJ PC6A screenings [3]

Complex	Results
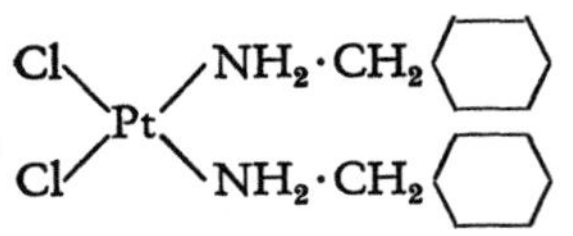 Cyclohexylmethylamine	Active T. I. = 2.6
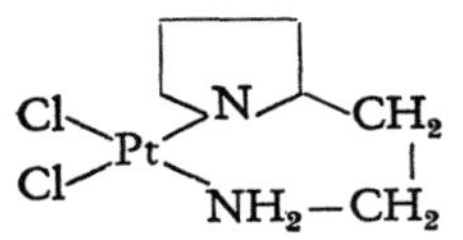 N-ethylaminopyrrolidine	Active T. I. = 11.1
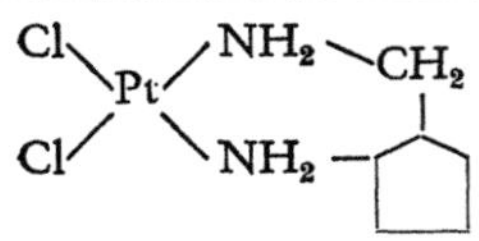 2-aminomethyl-1-*cyclo*pentylamine	Active T. I. = 40.7
*Iso*propylamine	Active T. I. = 37.2
*Iso*butylamine	Active T. I. = 13.4
*Iso*amylamine	Active T. I. = >125

Table 7. Second- and first-order rate constants for the reaction *cis*-$[Pt(am)_2 DMSO Cl]^+ + Cl^- + H^+ \rightarrow$ *cis*-$[Pt(am) DMSO Cl_2] + amH^+$ obtained from the rate law, Rate = $(k_1 + k_2 [Cl^-])[Pt\ am_2\ DMSO\ Cl^+]$, in methanol at 30.0°, $[H^+]$ = 0.01 M, ionic strength = 0.5

am	Ring size	pK_a	$10^5\ k_1\ s^{-1}$	$10^3\ k_2\ M^{-1}\ s^{-1}$
ammonia	—	9.25 [a]	2.6	1.04
cyclopropylamine	3	8.66 [b]	5.0	14.0
cyclobutylamine	4	9.34 [b]	0	3.17
cyclopentylamine	5	9.95 [b]	1.0	1.22
cyclohexylamine	6	9.82 [c], 9.84 [b]	0	1.16
cycloheptylamine	7	9.99 [c]	0	1.37
cyclooctylamine	8	10.01 [c]	0	1.37

[a] Measured in aqueous solution.

[b] Half-point scale reading in 50% water-ethanol (Roberts, J. D., Chambers, V. C.: J. Amer. Chem. Soc. 73, 5030 (1951)).

[c] Measured in 80% methylcellosolve-water (Prelog, V., Fausy, M., Neweihy, E., Häfliger, O.: Helv. Chim. Acta 33, 365 (1950)).

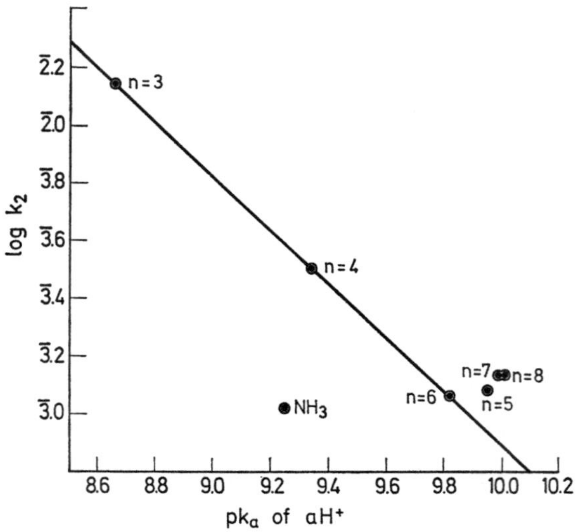

Fig. 1. Relationship between log k_2 and pk_a of aH^+ for *cis*-$[Pta_2(DMSO)Cl]^+ + Cl^- \rightleftharpoons$ *cis*-$[Pt(a)(DMSO)Cl_2] + a$

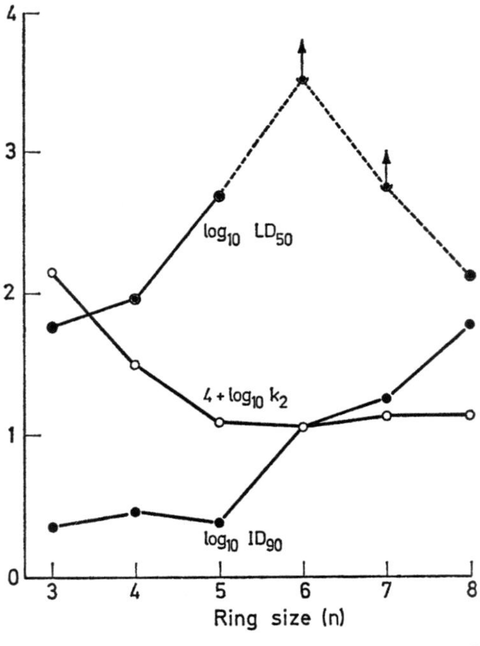

Fig. 2. A comparison of toxicity, potency and amine lability for *cis*-$[Pta_2Cl_2]$

It has been observed previously that guanosine reacts with $K_2[PtCl_4]$ to give a 2 : 1 complex, while adenosine gives a 1 : 1 complex [1, 4]. This author (rapporteur) suggests that these nucleoside complexes should be examined for anti-tumour activity. There has been recent speculation that the combination of a known anti-tumour drug

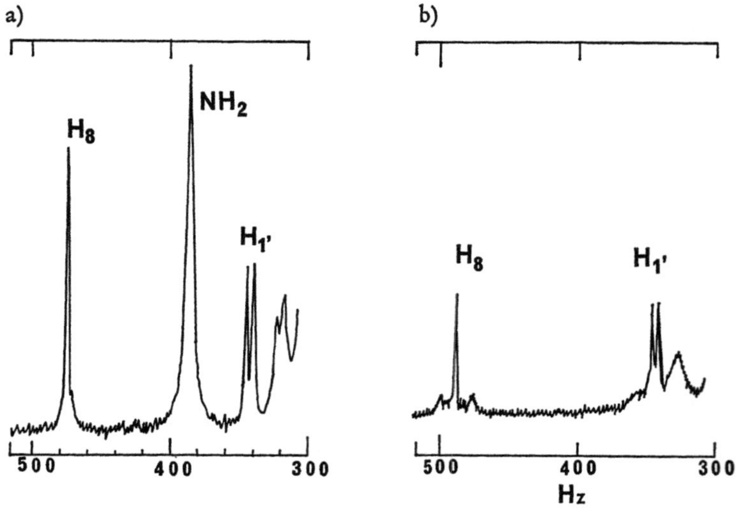

Fig. 3. a) N.m.r. spectrum of guanosine in DMSO; b) N.m.r. spectrum of [Pt(en)-(guanosine)$_2$]Cl_2 in D_2O

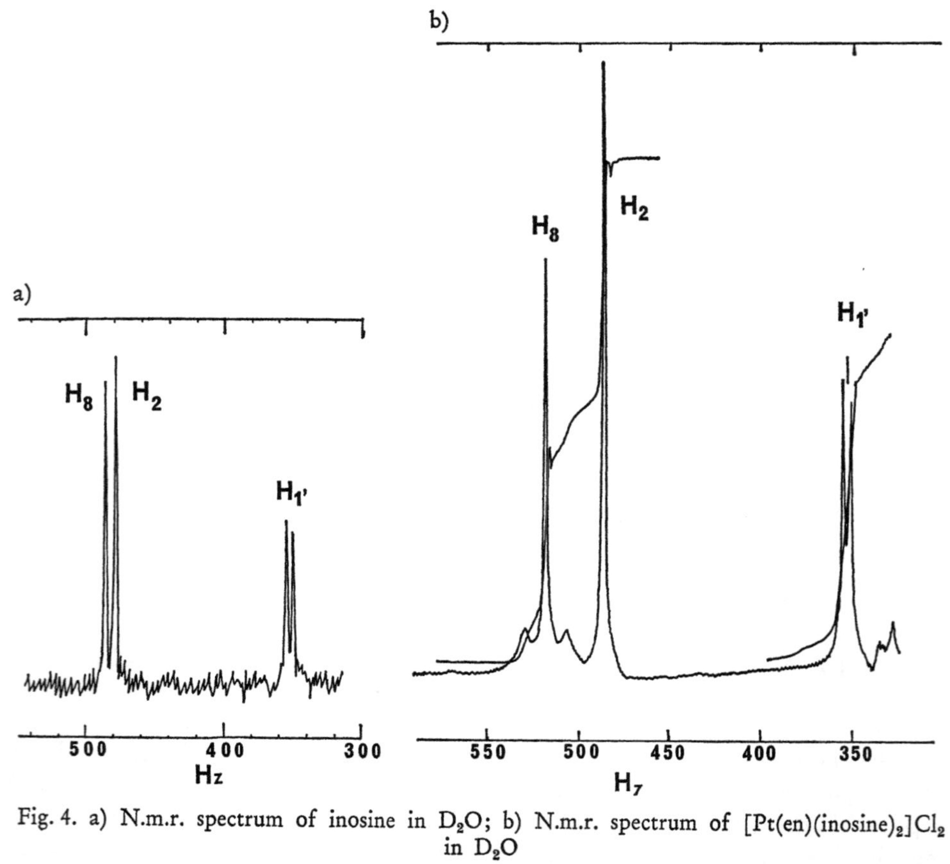

Fig. 4. a) N.m.r. spectrum of inosine in D_2O; b) N.m.r. spectrum of [Pt(en)(inosine)$_2$]Cl_2 in D_2O

with DNA could enhance its activity and selectivity. It is possible that this could also apply to nucleoside and particularly polynucleoside platinum complexes. Complexes with large molecules as ligands will not diffuse across the cell membrane. They require pinocytosis for uptake, after which lysozyme digestion may release the active portion of the complex. Tumour cells often have pinocytic, lysozomal and mitotic activity which is higher than normal cells and may therefore take up the drug more quickly than normal cells and thus be selectively affected. This technique has been successfully applied to daunorubicin — an anti-leukemia drug [5]. The activity of a blue platinum uracil complex, recently reported by MANSY and ROSENBERG (see Session II) may be similarly explained.

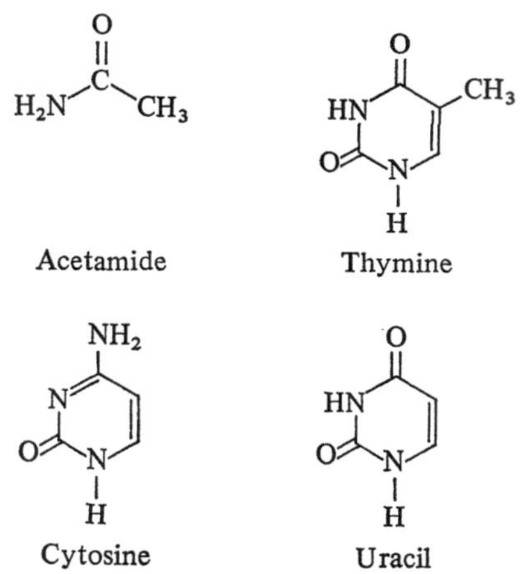

Fig. 5

Paper 3 (A. J. THOMSON, I. A. G. ROOS) discusses blue complexes of platinum. Their research is at an early stage and this paper compares some of the older platinum blue complexes with those which are now emerging from the reactions of DNA bases with platinum amine complexes. The majority of platinum blues are thought to be polymeric with short Pt-Pt distances but one group appears to be monomeric. These have acetamide or substituted acetamide ligands; the original complex (platinblau) was prepared from bis(acetonitrile)dichloroplatinum(II) and silver sulphate in water [6]. It is monomeric and was originally formulated as $[Pt(CH_3CONH_2)_2]\cdot H_2O$ from analytical data. The structure of these amido species is still rather uncertain; the latest formulation is $[Pt(IV)(amide\ anion)_2Cl_2]$ (or the corresponding hydroxide) However, other studies (particularly ESCA) strongly suggest that Pt(II) is present; a delocalised system is quite possible. Despite these structural problems the authors report some interesting observations:

1) *Cis*-$[Pt(NH_3)_2Cl_2]$ gives a certain amount of blue complex when dissolved in molten acetamide, whereas *trans*-$[Pt(NH_3)_2Cl_2]$, $[Pt(NH_3)_3Cl]Cl$, $[Pt(NH_3)_4]Cl_2$ and $[Pt\ en\ Cl_2]$ do not.

2) If crystals of $[Pt(CH_3CN)_2Cl_2]$ are left to stand in solutions of 1-methylthymine, 1-methylcytosine, uridine or DNA, a blue solution forms within a few hours (most readily with DNA). No blue solution is formed with 9-methyladenine or 9-methylguanine (the latter gives a creamy white precipitate).

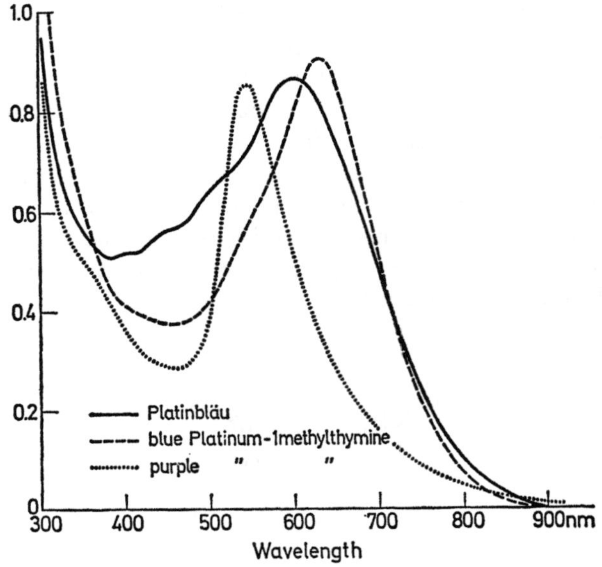

Fig. 6. U.V./visible spectra of some blue platinum species

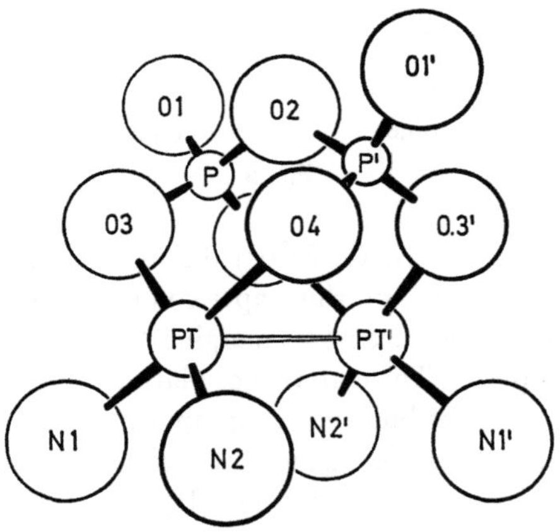

Fig. 7. Molecular structure of $Pt_2(NH_3)_4P_2O_7$

Fig. 5 shows some structural similarity between acetamide and the DNA bases. The blue colour is unlikely to be due to hydrolysis of acetonitrile as it does not occur with all of the bases. Moreover the reaction with DNA is particularly rapid. The reaction of *cis*-[$Pt(NH_3)_2Cl_2$] with nucleic acids is slow and does not occur in 0.1 N NaCl solutions. Thus with DNA the blue colour is likely to be due to co-ordination of thymine or a similar entity. *Cis*-[Pt(cyclopropylamine)$_2$Cl$_2$] on standing with 1-methylthymine (1 : 1) gives a purple complex. The infra-red spectrum suggests the presence of the amine and 1-methylthymine and two co-ordinated chlorides. Fig. 6

shows the vis/uv spectrum of this purple species compared with those of Platinblau and the blue complex obtained from reaction of 1-methylthymine with $[Pt(CH_3CN)_2Cl_2]$. Studies (mass spectra, NMR) aimed at elucidating the nature of these species are continuing. Samples have been sent for anti-tumour screening.

Paper 4 (Dr. J. A. STANKO) describes the structure and properties of *cis*-$[(Pt(NH_3)_2)_2P_2O_7]$ which is made by the action of inorganic pyrophosphate on *cis*-$[Pt(NH_3)_2Cl_2]$. Very few pyrophosphate complexes have been characterised and the structure of this one is depicted in Fig. 7. The major features are:

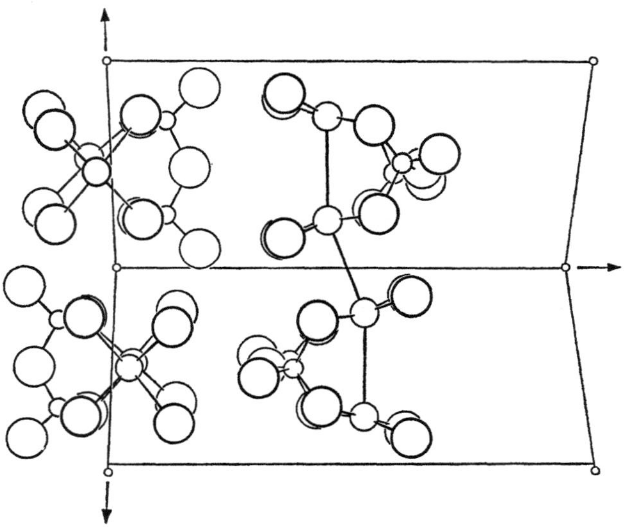

Fig. 8. Lattice arrangement of $Pt_2(NH_3)_4P_2O_7$

i) $P_2O_7^{4-}$ is acting as a tetradentate ligand using its four negatively charged oxygen atoms to co-ordinate in *cis* positions.

ii) The 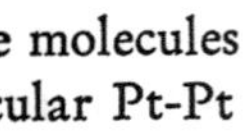 units are closely square planar.

iii) The adopted mode of chelation yields six membered 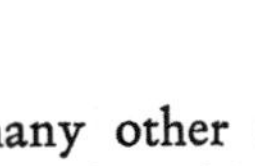 rings in a chair configuration.

iv) The Pt-Pt distance is 3.22 Å which implies some interaction between the platinum atoms.

v) The molecules are knit together in the crystal through a combination of short intermolecular Pt-Pt bonds (3.11 Å, shorter than intra-) and N-H...O = P hydrogen bonding (Fig. 8).

The implications of these structural results include:

a) The binding of *cis*-$Pt(NH_3)_2$ units (and possibly many other *cis*-Pta_2-) to pyrophosphate is stereochemically favourable. This ligand unit is found in ADP, TPP, NADH, FAD (etc.) and 1 : 1 and 2 : 1 complexes of ATP with $[PtenCl_2]$ have been noted previously [7].

b) Although nucleic acids only contain monophosphodiester links, the possibility of phosphate binding should not be neglected.

c) The importance of Pt-Pt bonding to overall stability is emphasised.

The refinement of the structure is not yet as low as desirable (R = 0.08) but this is sufficient to lend credibility to the basic structure.

Paper 5 (J. IBALL, S. N. SCRIMGEOUR) reports on the crystal structure of $[Pt(en)Cl_2]$ (Fig. 9). The platinum adopts the usual square planar configuration with a Pt-Pt distance of 3.381 Å which compares with 3.372 and 3.409 Å reported for *cis*-$[Pt(NH_3)_2Cl_2]$ and 3.45 Å for $[Pt(bipy)Cl_2]$. The Cl-Cl distance (important for considerations of possible binding sites) is 3.4 Å. The $[Pd(en)Cl_2]$ analogue is under study; the unit cell dimensions of the two compounds are almost identical..

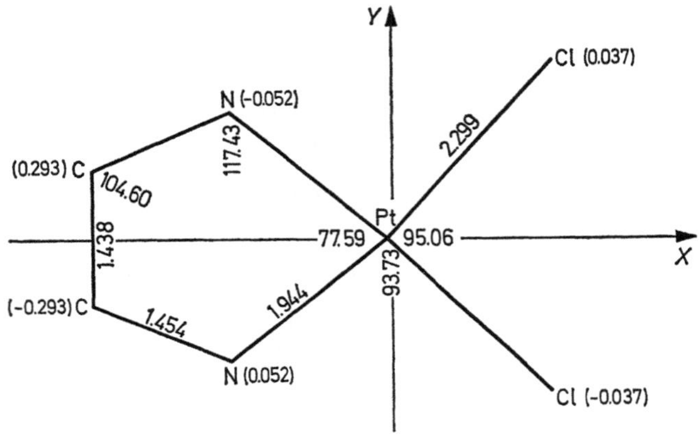

Fig. 9. Crystal structure of $[Pt(en)Cl_2]$

III. Platinum(IV)

Relatively little work has been carried out on Pt(IV) complexes since the initial discovery [8] of activity in *cis*-$[Pt(NH_3)_2Cl_4]$ and $[Pt(en)Cl_4]$. It is likely — but not proven — that their activity is due to *in vivo* reduction to the corresponding Pt(II) species. This possibility might be invalidated if one could prepare an active Pt(IV) species which is reduced to an inactive Pt(II) species. Pt(IV) compounds may be of interest as anti-tumour agents as their membrane transport properties (prior to reduction) may differ from Pt(II) compounds [9].

Paper 6 (J. V. RUND) discusses oxidative addition reactions of Pt(II) starting with dihalogeno-phenanthroline Pt(II) complexes. Although the chloro complex has shown relatively little anti-tumour activity, these oxidation addition reactions provide preparative routes for many Pt(IV) complexes.

i) Symmetric addition using halogens and thiocyanogen. Iodine (in excess) gives species which appear to contain both iodide and triiodide ligands e. g. $[Pt(phen)I_3(I_3)]$, $[Pt(phen)Br_2I(I_3)]$.

ii) Unsymmetric additions: The addition takes place in two steps; these can be achieved by using a halogen in the presence of other halide ions (e. g. Br_2 in HCl but not I_2 in HCl). Some mixed halogens can also be used, e. g. ICl, ICN but BrCN was unsuccessful. Nitrosyl chloride tends to act as a chlorinating agent, although $[Pt(phen)I_2(NO)Cl]$ can be isolated from ice-cold chloroform ($\nu NO = 1664\ cm^{-1}$).

IV. Other Metals

Since the demonstration of potent anti-cancer activity for certain platinum complexes, an intriguing question has been "How general is this property for other transition metals". Table 8 attempts to summarise the relevant findings prior to this Conference [10]. Starting from the elementary supposition that for anti-tumour activity a complex should be of intermediate lability, fundamental co-ordination chemistry can predict the metal systems which are most likely to possess activity, as discussed in the opening paper (Dr. R. J. P. WILLIAMS).

Table 8. Reported anti-tumour activity in metals other than platinum

Metal ion		Anti-tumour activity	Type of complex	Bacterial effects
Co(III)	d^6(L)	±	$[Co(NH_3)_4Cl_2]^+$ [a]	
Cr(III)	d^3	±	$[Cr(NH_3)_4(C_2O_4)]^+$ [a]	
Ni(II)	d^8	+	Butylphthalate chelate [b]	
Cu(II)	d^9	+	Oxime chelates [c] thiosemicarbazones [d, e, f]	
Rh(III)	d^6(L)	+	Amines [g]	+ [h]
Ir(III)	d^6(L)	−	Amines [g]	± [i, g, h]
Ir(III)/Ir(IV)	d^6/d^5	+	uv-irradiated [j] $(NH_4)_2[IrCl_6]$	+ [j]
Pd(II)	d^8	−	Amines [k, l]	Bacteriostatic [l]
Ru(III)	d^5(L)	(−)	$Ru(NH_3)_3Cl_3$ [m]	+ [i, g]

[a] KAJIWARA, 1951. [b] GESCHICKTER and REID, 1947. [c] TAKAMIYA, 1960. [d] CRUN and PETERING, 1967. [e] BOOTH and SARTORELLI, 1967. [f] CHARLSON, Paper 10. [g] CLEARE, CONNORS, HOESCHELE, ROSENBERG and VAN CAMP, Paper 9. [h] GILLARD, HARRISON and MATHER, Paper 8. [i] BECK and ROSENBERG, 19. [j] GALE, WALKER, SMITH and STONE, 1970, 1971. [k] CLEARE and HOESCHELE, 1973. [l] GRAHAM and WILLIAMS, Paper 7. [m] CLEARE, HOESCHELE, VAN CAMP and CONNORS, 1973.

A. Palladium

Studies on Pd(II) analogues of active Pt(II) complexes have reported only marginal activity against the S. 180 tumour (Table 9) [1]. *Paper 7* (R. D. GRAHAM, D. R. WILLIAMS) report on the effects of a number of chloro, chloroamine and amino-acid complexes of Pd(II) and Pd(IV) on various biological systems (Table 10). The compounds showed little activity against a wide range of viruses. However, *cis*-$[Pd(NH_3)_2Cl_4]$ showed some anti-tumour activity which is somewhat surprising as the corresponding *cis*-Pd(II) complex appears to be inactive, and Pd(IV) is readily reduced to Pd(II) (more readily than Pt(IV) → Pt(II)). Pd(IV) is a d^6 system and may last

Table 9. Animal testing results for palladium(II) complexes [a] [1]

Complex	Solvent	Dose range (mg/kg)	Dose response	Toxic level (mg/kg)	T/C	Dose (mg/kg)
cis-$[Pd(NH_3)_2Cl_2]$	S	1.25— 10	—	> 10	83	1.25— 10
$[Pd(en)Cl_2]$	SS	5 — 40	—	> 40	79	5 — 40
$[Pd(en)Cl_2]$	SS	10 — 50 [b]	—	> 50	78	10 — 50
$[Pd(NH_3)_2mal]$	WS	25 —200	±	150—200	55	75 —150
$[Pd(en)mal]$	W	25 —100	—	>100	75	25 —100
$[Pd(en)ox]$	W	1.25— 75	—	> 40	104	1.25— 25
$[Pd(NH_3)_2ox]$	SS	1.25—200	—	~100	80	12.5 — 80

[a] Sarcoma 180 in Swiss white mice.
[b] 4 injections/day for 2 days.

Table 10. Biological screening of palladium complexes (Graham and Williams, paper 7)

	Cytotoxicity	Antitumour	Antiviral [a]	Fungicidal
c-$[Pd(NH_3)_2Cl_2]$	—		±	
t-$[Pd(NH_3)_2Cl_2]$		— [b]	—	—
c-$[Pd(NH_3)_2Cl_4]$	—	+ [c]	—	—
$[PdenCl_2]$	—		—	
$Na_2[PdCl_6]$	—		—	
$(NH_4)_2[PdCl_4]$	—		±	
$[Pd(NH_3)_4][PdCl_4]$	—		±	
$(enH_2)[PdCl_4]$	—		—	
$[(\pi\text{-}C_3H_5)PdCl]_2$	+		—	

[a] ± indicates incomplete inhibition of >1 virus.
[b] reduces S 180 and Landschutz ascites tumours by 45% (50 mg/kg-daily).
[c] reduces S 180 and Landschutz ascites tumours by 82% (100 mg/kg-daily).

long enough to enjoy better membrane transport or may not react so quickly in the blood as the Pd(II) derivative. The authors consider that the most promising compounds are *cis*-$[Pd(NH_3)_2Cl_2]$, *cis*-$[Pd(NH_3)_2Cl_4]$ and $[\pi\text{-}C_3H_5PdCl]_2$. The first two are analogous to active platinum compounds, but the π-allyl complex contains neither a N-H group or *cis* leaving chlorides (although break up of the chloride bridge will have a similar result). These authors caution against an overemphasis on platinum compounds. Although testing of Pd(II) complexes to toxic limits and on different dose schedules might reveal more activity, it is likely that Pd(II) does not possess effective anti-tumour activity due to its greater reactivity ($\sim 10^5$ higher than Pt(II) for comparable compounds) [1]. Au(III) complexes are also relatively reactive (of the order of 10^4 greater than platinum) but no screening results have been submitted to this Conference or reported elsewhere.

B. Rhodium(III)

There are two papers which discuss biological properties of Rh(III) complexes. *Paper 8* (R. D. Gillard, K. Harrison, I. M. Mather) reports studies on the inhibition of growth and cell division of *E. coli* by some fifty Rh(III) complexes with N-donor ligands most of which are of the type *trans*-$[RhL_4X_2]Y$, where $L_4 = 4$ pyridine or pyridine type monodentate ligands or 2 bidentate N donors such as ethylenediamines, X = Cl, Br, Y = Cl, Br, NO_3, ClO_4. The active complexes can either inhibit growth or, at lower concentrations which permit significant growth, cell division, and hence cause filament formation.

Table 11. Complexes of Rh(III) with other nitrogen-containing heterocycles

Complex	Filamentation pattern	$E_{1/2}$	log K_D
A. N-methylimidazoles			
$[Rh(miz)_5Cl](ClO_4)_2$	inactive	−420	1.66
$[Rh(miz)_5Br](ClO_4)_2$	inactive	−370	
t-$[Rh(miz)_4Cl_2]Cl$	inactive	−120	1.85
t-$[Rh(miz)_4Br_2]Br$	30% 3—20×	− 90	1.76
t-$[Rh(5\text{-}Cl\text{-}miz)_4Cl_2]Cl$	30% 3—30×	−260	0.77
t-$[Rh(5\text{-}NO_2\text{-}miz)_4Cl_2]Cl$	45% 3—30×	− 70	0.58
B. N-methylimidazole, pyrazole and thiazole			
t-$[Rh(miz)_4Cl_2]Cl$	inactive	−120	1.85
t-$[Rh(pzl)_4Cl_2]Cl$	10% 3—20×	0	0.43
t-$[Rh(thz)_4Cl_2]Cl$	100% 20—30×+	+ 90	1.94

$$K_D = \frac{\text{concentration of complex in aqueous phosphate buffer (pH 7)}}{\text{concentration of complex in octanol}}$$

miz = N-methylimidazole.
pzl = pyrazole.
thz = thiazole.

The major results are:

i) Most complexes exhibiting activity (however slight) have reduction potentials more positive than −190 mV with respect to a standard hydrogen electrode. With a wide range of $E_{1/2}$ only the most easily reduced compounds show activity (Tables 11 and 12; there are many other examples).

ii) When the $E_{1/2}$ values are all close to zero, the ability to force filamentous growth appears to be somewhat related to lipophilic character (Table 13). Compounds with lower and negative log K_D's are more effective. The apparently lower activity of compounds with bulky substituents on pyridine could be due to either steric or $E_{1/2}$ effects.

iii) Neither the position of substitution nor the counterion are important.

iv) For the *trans* ligands (X), bromide and iodide appear to be better than chloride but again this could be due to $E_{1/2}$ or lipophilic effects (Table 14).

In these complexes the Rh(I) state is generated comparatively easily and is thought (in catalytic quantities) to be responsible for the enhanced chemical reactivity

Table 12. Complexes of Rh(III) with substituted ethylenediamines

Complex	Filamentation pattern	$E_{1/2}$	log K_D
c-$[Rh(en)_2Cl_2]NO_3$	inactive	−520	∞
t-$[Rh(en)_2Cl_2]NO_3$	inactive	−460	1.78
c-$[Rh(udmen)_2Cl_3]ClO_4$	inactive	−420	2.19
t-$[Rh(udmen)_2Cl_2]ClO_4$	inactive	−230	1.68
t-$[Rh(sdmen)_2Cl_2]Cl$	inactive	−290	2.05
t-$[Rh(trimen)_2Cl_2]ClO_4$	3% 5×	−240	2.21
t-$[Rh(tetmen)_2Cl_2]Cl$	100% 60×+	−110	2.31

en = ethylenediamine.
udmen = *N,N*-dimethylethylenediamine.
sdmen = *N,N'*-dimethylethylenediamine.
trimen = *N,N,N'*-trimethylethylenediamine.
tetmen = *N,N,N',N'*-tetramethylethylenediamine.

Table 13. Change in activity on substitution of pyridine in *trans*-$[Rh(py)_4Cl_2]Cl$

Complex	Filamentation pattern	$E_{1/2}$	log K_D
t-$[Rh(py)_4Cl_2]Cl$	85% 10—30×*	+ 30	1.71
t-$[Rh(4\text{-}Mepy)_4Cl_2]Cl$	98% 20—100×*	−150	0.68
t-$[Rh(3\text{-}Etpy)_4Cl_2]Cl$	98% 50—100×*	− 30	−0.19
t-$[Rh(iquin)_4Cl_2]Cl$	100% 100—300×*	+ 20	−2.12
t-$[Rh(4\text{-}nPrpy)_4Cl_2]Cl$	50% 50—100×*	−120	−2.24
t-$[Rh(4\text{-}tBtpy)_4Cl_2]Cl$	25% 3—30×*	−190	−2.70

Table 14. Change in activity on variation of X in *trans*-$[Rh(py)_4X_2]Y$

Complex	Filamentation pattern	$E_{1/2}$	log K_D
t-$[Rh(py)_4I_2]I$	98% 20—100×*	− 10	0.88
t-$[Rh(py)_4Br_2]Br$	98% 20—60×*	− 30	1.20
t-$[Rh(py)_4Cl_2]Cl$	85% 10—30×* 100% 20—50×+	+ 30	1.71
t-$[Rh(py)_4Cl(OH)]ClO_4$	n—50×	−100	2.02
t-$[Rh(py)_4(N_3)_2]N_3$	n—30×	− 80	1.76

py = pyridine.
iquin = isoquinoline.
Me = methyl, Et = ethyl, nPr = n-propyl, tBt = tertiary butyl.
* Screened at 10^{-5} M.

Table 15. Comparison of the biological activities of complexes containing rhodium(III) and complexes of platinum

Complex	Concentration (M)	Effect on growth of E. coli
t-$[Rh(py)_4Cl_2]Cl$	5×10^{-5}	100% 20—50× +
t-$[Rh(iquin)_4Cl_2]Cl$	10^{-5}	100% 100—300×
c-$[Pt(NH_3)_2Cl_2]$	5×10^{-5}	98% 150×
$[Pt(bipy)Cl_2]$	5×10^{-5}	No growth
	5×10^{-6}	10% 3—30×
$[Pt(en)(mal)]$ ǂ	5×10^{-5}	100% 20—100× +
$[Pt(bipy)_2](NO_3)_2$	1.5×10^{-5}	No growth within 24 hours
	10^{-5}	Slightly inhibited growth. Cells are swollen
$[Pt(bipy)(NH_3)_2]Cl_2$	5×10^{-5}	20% 3—30×
$[Pt(py)_4](BF_4)_2$	5×10^{-5}	Normal growth
$[Pt(NH_3)_4](ClO_4)_2$	5×10^{-5}	Normal growth
$[Pt(py)_4Br_2](BF_4)_2$	5×10^{-5}	n—150×. Growth inhibited Filaments are swollen
$[Pt(NH_3)_4Cl_2]Cl_2$	5×10^{-5}	Normal growth
$[Pt(dien)Cl]Cl$	5×10^{-5}	15% 3—20×
$[Pt(dien)I]I$	5×10^{-5}	5% 3—10×
t-$[Pt(3,5\text{-diMepy})(CO)Cl_2]$	10^{-5}	No growth within 24 hours
	5×10^{-6}	Normal growth
t-$[Pt(4\text{-CNpy})(C_2H_4)Cl_2]$	5×10^{-6}	No growth within 24 hours
	5×10^{-7}	Very inhibited growth. Slight filamentation.
$K_2[PtCl_4]$	5×10^{-5}	No growth within 24 hours
	10^{-5}	10% 3—30× +
1,2,4-$[Rh(NH_3)_3Cl_3]$ ǂ	5×10^{-5}	15% 3—20×
1,2,4-$[RuNH_3)_3Cl_3]$ ǂ	5×10^{-5}	100% 30—100×
1,2,4-$[Ir(NH_3)_3Cl_3]$ ǂ	5×10^{-5}	3% 5×
1,2,4-$[Rh(py)_3Cl_3]$	5×10^{-5}	15% 3—20×

bipy = 2,2′-bipyridyl.
mal = malonate anion.
dien = diethylenetriamine.
ǂ Complexes supplied by Dr. M. J. CLEARE.

of many of the compounds (i. e. those with high $E_{1/2}$). Rh(I) in aqueous solution is greatly inhibited by oxygen, but favoured by addition of 2 electron reducing agents. The effect of Rh(I) was studied:

a) by growing *E. coli B* anaerobically in the presence of *trans*-$[Rh(4\text{-n-}Prpy)_4(Cl)(OH)]^+$ (n-Pr = n-propyl).

b) by growing *E. coli B* aerobically with some air stable Rh(I) complexes.

In both cases bacteriostasis occurred at a comparable concentration to that for the Rh(III) measurements.

With some as yet uncharacterised Rh(I) complexes slow growth occurred giving swollen distorted organisms with ruptured cell walls [1]. There was little effect on growth at low concentrations. Thus the activity of the Rh(III) complexes is not

[1] The reaction of $[Rh(CO)_2Cl]_2$ in DMF with ligands such as o-phen gives green Rh(I) species which are being characterised.

Table 16. Screening of rhodium amine complexes (I)

Sarcoma 180					
Complex	Dose range	Dose response	Toxic level	T/C	Dose
m-$[Rh(NH_3)_3Cl_3]$(S.S.)	12—100	+	70—100	17	100 [a]
(S)	5— 30 [b]	+	~ 20	31	20
m-$[Rh(dien)Cl_3]$(S.S.)	25—100	±	>100	54	50—100
t-$[Rh(NH_3)_3Cl_2(H_2O)]NO_3$(W)	50—150	±	125—150	48	150
m-$[Rh(NH_3)_3(Cl)ox]$(W.S.)	50—125	—	>125	73	125
t-$[Rh(NH_3)_4Cl_2]NO_3$(W)	25—100	—	>100	82	25—100
t-$[Rh(py)_4Cl_2]Cl$(S)	20— 50	±	40—50	45	50 [a]
m-$[Rhpy_3Cl_3]$(S.S.)	50—200	±	~100	51	100 [a]
c-$[Rh(en)_2Cl_2]NO_3$(S)	50—200	±	>200	63	50—200
t-$[Rh(en)_2Cl_2]NO_3$(S)	5— 50	±	> 50	46	50

[a] 66% survivors only.
[b] daily dose for 8 days.
ox = oxalate.

Table 17. Screening of rhodium complexes (II)

Sarcoma 180 (ascites) [a]					
	Dose range	Dose response	Toxic level	I.L.S. [d]	Dose
$[Rh(NH_3)_3Cl_3]$(S.S.)	50— 100	+	>100	23%	100
ADJ/PC6A [b]					
	Dose range	Dose response	LD_{50}	ID_{90}	T.I.
$[Rh(NH_3)_3Cl_3]$(A)	4— 500	+	235	86	2.6
$[Rh(NH_3)_3(NO_3)_3]$(A)	12—1500	+	135	59	2.3
Walker 256 [c] carcinosarcoma					
$[Rh(NH_3)_3Cl_3]$	5— 160	+	~170	~40	~4

[a] Swiss white mice.
[b] BALB/c̄ mice.
[c] Rats.
[d] % increase in life span.

simply a function of [Rh(I)] in the growth medium. The authors suggest that Rh(I) may be generated within the cell envelope, possibly on the membrane where the reductive enzymes of the electron transport chain are situated.

The primary site of action of rhodium compounds remains in doubt. The Rh(III) complexes do not appear to interact to any appreciable extent with DNA as there is little effect on DNA synthesis, and they are poor inducers of phage from a lysogenic bacterium. Finally the paper compares the effects on bacteria of Rh(III) and Pt(II) complexes and the rhodium and iridium triammines which are discussed in Paper 9 (Table 15). The superficial effects of the various complexes are very similar, i. e. morphological changes, depression of oxygen uptake and the lack of inhibition of DNA synthesis. However, more extensive biological studies may well reveal significant differences in their mechanism of action (cf. penicillin — causes filamentation by inhibiting cell wall synthesis).

The authors point out that some charged Pt(II) and Pt(IV) complexes show bacterial activity and contrast this with the empirical rules which were discussed earlier with regard to anti-tumour activity of Pt(II) complexes. In this regard the rapporteur would like to point out that although some correlation between bacterial and anti-tumour activity exists, it is not a direct relationship. All anti-tumour Pt(II) complexes cause filamentation but the reverse is far from true. The charged Pt(II) species which cause filamentation generally do so to a lesser extent than do the active anti-tumour compounds; apart from $[Pt(py)_4Br_2](BF_4)_2$ the results shown in Table 15 support this view.

Paper 9 (M. J. Cleare, T. A. Connors, J. D. Hoeschele, L. Van Camp, B. Rosenberg) report on some anti-tumour screening results for Rh(III) and Ir(III) amine complexes. The compounds were chosen with the platinum data in mind i. e. neutral complexes with *cis* leaving groups. The neutral trichlorotriamminerhodium(III) species displays definite activity against four tumour systems (Tables 16 and 17). The tridentate amine complex $[Rh(dien)Cl_3]$ shows marginal activity against the S. 180 tumour as does the dichloroaquotriammine salt and both are undergoing further tests.

Table 18. Screening of iridium amine complexes

Sarcoma 180 [a]					
	Dose range	Dose response	Toxic level	T/C	Dose
$[Ir(NH_3)_3Cl_3]$(W.S.)	50— 200	—	150—200	76	200 [b]
c-$[Ir(en)_2Cl_2]NO_3$(S)	50— 200	—	> 200	94	50—200
t-$[Ir(en)_2Cl_2]NO_3$(S)	50— 200	—	> 200	109	50—200
$Cs[Ir(NH_3)_2Cl_4]$	50— 200	—	> 200	111	50—200
$Cs[Ir(en)Cl_4]$	50— 200	—	> 200	95	50—200
t-$[Ir(NH_3)_4Cl_2]Cl$(S)	50— 200	—	> 200	75	50—200
Sarcoma 180 (ascites) [a]					
$[Ir(NH_3)_3Cl_3]$(S.S.)	125— 200	—	> 200	— 17% [c]	125—200
ADJ/PC6A [d]					
			LD_{50}	ID_{90}	T.I.
$[Ir(NH_3)_3Cl_3]$(A)	12—1500	—	>1500	—	—

[a] Swiss white mice.
[b] 66% survivors.
[c] % I.L.S.
[d] BALB/c̄ mice.

The trichlorotriammine derivative shows activity inferior to the *cis*-Pt(II) drugs except against the resistant Walker tumour. The results again suggest that charged complexes are generally inactive. The triammine aquo species was made in order to solubilise the triammine in the hope that reconversion will occur via chloride within the animal's body. The *trans* tetrapyridine complex whose bacterial effects have been discussed in the previous paper showed marginal activity against the S. 180 tumour. Gillard and Connors report that initial tests of some of the other compounds in Paper 7 have not given promising results although further screening is continuing.

This again suggests a difference between the *trans*-pyridine type rhodium complexes and the platinum(II) amines. Again the anti-tumour active rhodium(III) compounds caused filamentation, but all the compounds which caused some degree of filamentation were *not* anti-tumour active.

None of the iridium complexes showed any anti-tumour activity against S. 180 at the levels tested (Table 18). The toxic levels (if reached) are very high and as d^6 complexes are always fairly inert (especially for a 3rd row metal) it seems likely that they are too unreactive to be metabolised prior to excretion.

C. Metal Chelates

The final paper (No. 10 by A. J. Charlson) discusses the synthesis of a variety of metal chelates as anti-cancer substances. The chelates involved are thiosemicarbazones, oximes and amino acids.

i) Thiosemicarbazones. Guanylhydrazones and thiosemicarbazones of glyoxal and methyl glyoxal have anti-leukemic properties (Fig. 10). 3-Ethoxy-2-oxobutyraldehyde bis(thiosemicarbazone) — (KTS) — has attracted a lot of study and its anti-tumour activity has been found to be directly proportional to the dietary intake of Cu(II) ions. Cu(II) ions alone have no effect but the Cu-KTS chelate was equally (if not more) effective but was more toxic. This suggests that KTS acts by bringing copper

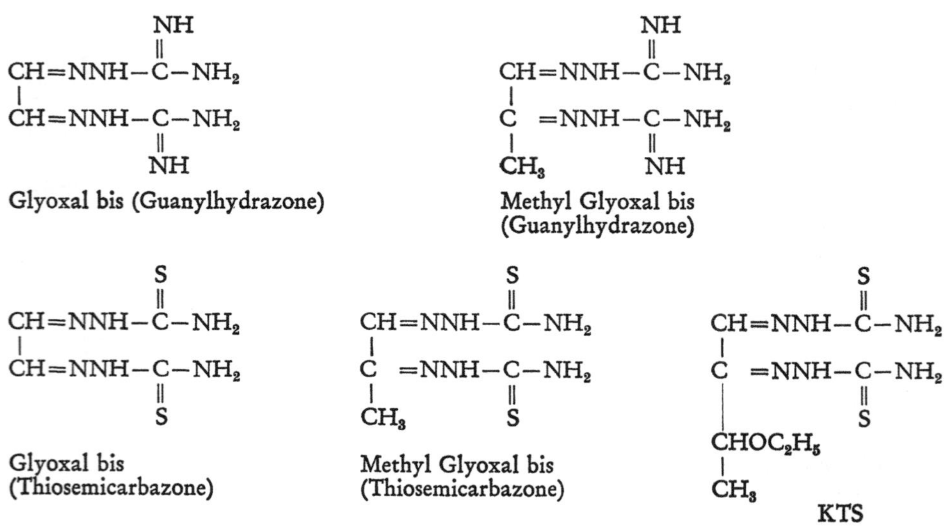

Glyoxal bis (Guanylhydrazone)

Methyl Glyoxal bis (Guanylhydrazone)

Glyoxal bis (Thiosemicarbazone)

Methyl Glyoxal bis (Thiosemicarbazone)

KTS

Fig. 10

into the tumour cells. The author has synthesised a number of chelates of bis(thiosemicarbazones) of some sugar ozones. The probable structure is shown in Fig. 11 (for D-glucosone) and has square planar geometry. Table 19 shows the results so far obtained for the various chelates against leukemias L. 1210 and P. 388. The complexes

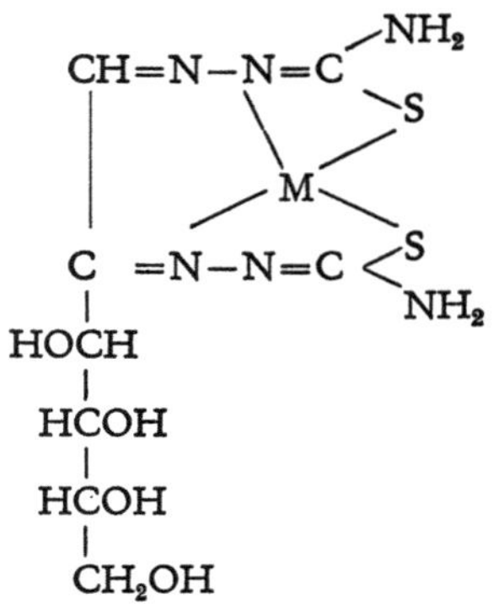

Fig. 11. Probable structure of D-glucosone-bis-thiosemicarbazone metal complexes

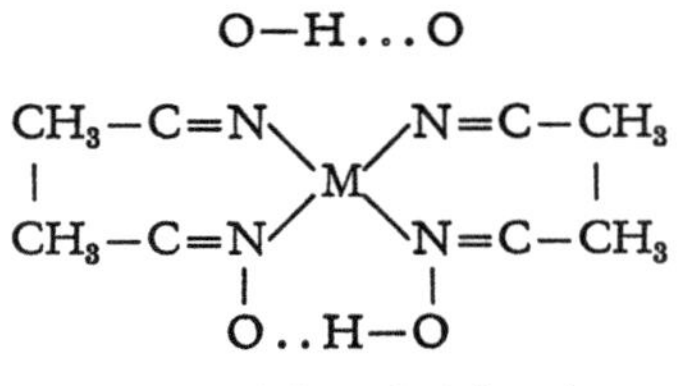

Fig. 12. The structure of dimethylglyoxime metal complexes

Table 19. *D*-Glucosone bis (Thiosemicarbazone)

Substance	Leukemia 1210	Dose level [a] mg/kg	Leukemia P 388	Dose level [a] mg/kg	KB Cell culture	ED_{50} [b]
Ligand	inactive	100	inactive	100	inactive	100
Copper chelate	inactive	25	inactive	40	*active*	2.8
Palladium chelate	inactive	200			inactive	8.6
Nickel chelate	inactive	400			inactive	29
Platinum chelate	inactive	200				
Iron chelate	inactive	200				
Rhodium chelate	inactive	400			inactive	100

[a] Dose levels higher than the amounts recorded were toxic in most of the test animals.
[b] Expressed in microgrammes per ml.

are all inactive although the copper chelate was cytotoxic to the cell culture while the ligand was not. Similar results are reported for L-Rhamnosone bis(thiosemicarbazone). The palladium chelate of methylglyoxal bis(thiosemicarbazone) was inactive and the palladium chelate of KTS is under test.

ii) Oximes. Oximes have been reported to show some activity in the presence of Cu(II) ions but not with other 1st row transition metals. The structure of dimethylglyoxime complexes is shown in Fig. 12. The copper, nickel and palladium chelates have been submitted to the NCI for screening. The results are not encouraging but the palladium complex shows cytotoxicity towards cells in culture (Table 20).

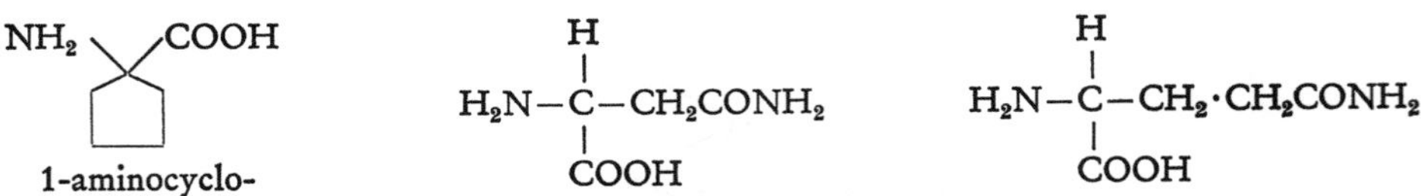

1-aminocyclopentane-carboxylic acid. L-Asparagine L-Glutamine

Fig. 13

Table 20. Dimethyl glyoxime

Substance	Leukemia 1210	Dose level mg/kg	KB Cell culture	ED_{50}
Copper chelate	inactive	12.5	inactive	20
Nickel chelate	inactive	25	inactive	9.4
Palladium chelate			active	0.93

iii) Amino Acids. This can be compared with the work of Williams and Graham — Paper 6. Those under study are shown in Fig. 13. 1-aminocyclopentanecarboxylic acid was considered because this ligand is known to inhibit tumours; the copper, zinc and palladium chelates have been prepared. The only compound tested, the copper complex, was inactive against the L. 1210 tumour. Copper and palladium complexes of L-glutamine and L-asparagine were prepared in the hope that they may be specifically taken up by tumours. In their paper Williams and Graham go further and postulate that the glutamine and serine complexes should be tested against myeloid leukemia and the asparagine complex against lymphatic leukemias.

References

1. Cleare, M. J., Hoeschele, J. D.: Bioinorganic Chemistry **2**, 187 (1973).
2. Connors, T. A., Jones, M., Ross, W. C. J., Braddock, P. D., Khokhar, A. R., Tobe, M. L.: Chem.-Biol. Interactions **5**, 415 (1972).
3. Connors, T. A., Jones, M., Braddock, P. D., Khokhar, A. R., Tobe, M. L.: (1973) — personal communication.
4. Cleare, M. J., Hoeschele, J. D.: (1973) — unpublished results.
5. Trouet, A., Deprez de Campeneere, D., De Duve, C.: Nature (New Biol.) **239**, 112 (1972).
6. Hoffman, K. A., Bugge, G.: Ber. **41**, 312 (1908).
7. Robins, A. B.: Chem.-Biol. Interactions **6**, 95 (1973).
8. Rosenberg, B., Van Camp, L., Trosko, J. E., Mansour, V. H.: Nature (Lond.) **222**, 385 (1969).

9. THOMSON, A. J., WILLIAMS, R. J. P., RESLOVA, S.: Structure and Bonding (Berlin) **11**, 1 (1972).
10. CLEARE, M. J.: Coord. Chemistry Reviews **12**, 349 (1974).
11. KAJIWARA, K.: Gann **42**, 272 (1951).
12. GESCHICKTER, C. F., REID, E. E.: In: Approaches to Tumour Chemotherapy (Ed. E. R. MOULTON), A.A.A.S., Washington, D.C. (1947) Pg. 431.
13. TAKAMIYA, K.: Nature (Lond.) **185**, 190 (1960).
14. CRUN, J. A., PETERING, H. G.: Cancer Res. **27**, 1278 (1967).
15. BOOTH, B. A., SARTORELLI, A. C.: Cancer Res. **27**, 1614 (1967).
16. GALE, G. R., WALKER, E. M., JR., SMITH, A. B., STONE, A. E.: Proc. Soc. exp. Biol. (N.V.) **136**, 1197 (1971); ibid. **135**, 690 (1970).
17. BECK, D., ROSENBERG, B.: Unpublished results.
18. CLEARE, M. J., HOESCHELE, J. D., VAN CAMP, L., CONNORS, T. A.: To be published.

The Interactions of Platinum Compounds with Biological Molecules

A. J. THOMSON

This article describes the modes of interaction of anti-tumour and related platinum complexes with biomacromolecules and their components. We start with a brief review of the chemistry of the platinum compounds and proceed to illustrate where possible this chemistry as it is seen to occur in the interaction of platinum compounds with biological molecules. Our aim is to uncover this chemistry as it may apply to the transport of platinum compounds in a biological medium, to the attack by the platinum drugs at their site of action or to their role in causing toxicity. Such studies are only able to indicate possible modes of action and can never furnish proof of a mechanism of biological activity. However, we shall refer most often to four prototype compounds, namely *cis*- and *trans*-dichloro-diammine platinum(II) and *cis*- and *trans*-tetrachloro-diammine platinum(IV), whose stereo-chemistries are shown in Fig. 1. Of these four compounds only the *cis* isomers display appreciable anti-tumour

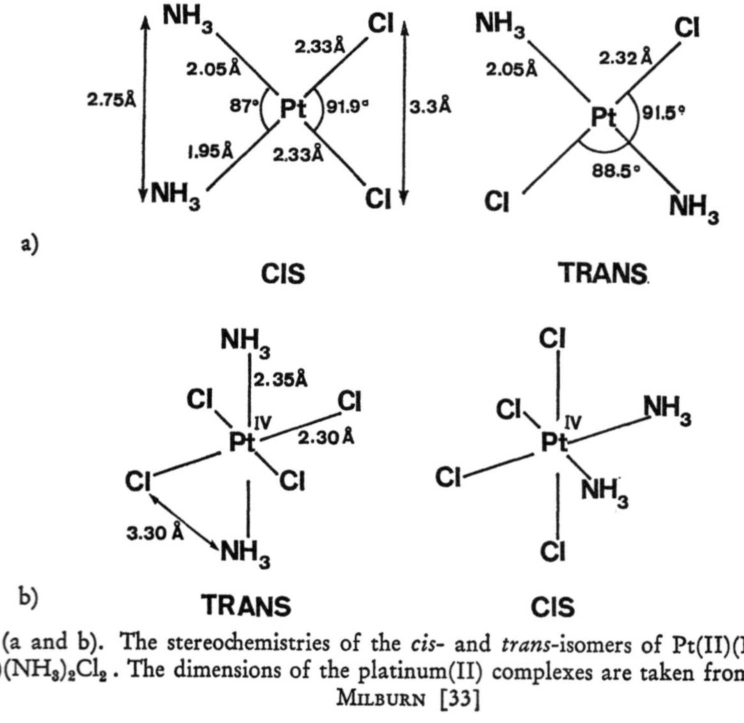

Fig. 1 (a and b). The stereochemistries of the *cis*- and *trans*-isomers of $Pt(II)(NH_3)_2Cl_2$ and $Pt(IV)(NH_3)_2Cl_2$. The dimensions of the platinum(II) complexes are taken from TRUTER and MILBURN [33]

activity, the *trans* isomers being largely ineffective. Thus a comparison of the reactions of these four complexes with biological molecules can provide an invaluable index of the relevance or otherwise of the chemistry to the biological effect. Therefore although the *trans* isomers are ineffective as anti-tumour drugs their reactions with macro-molecules are of great interest and comparative studies should always be made where possible.

Chemical Features of Platinum Drugs

The chemical reactivity of platinum and related metal complexes has been described in detail by [32] and here we summarise their conclusions for the case of platinum complexes, treating first the planar platinum(II) compounds followed by the octahedral platinum(IV) complexes.

Cis- and *Trans*-Pt(II)$(NH_3)_2Cl_2$

a) Substitution Reactions. Both isomers, in aqueous solution, exchange only two of their ligands, the chloride ions, for incoming groups such as H_2O, OH^-, R—NH_2, RS^- or R—S—CH_3, which are abundant in a biological milieu. For kinetic and thermodynamic reasons the amine groups are difficult to replace (except in the special case mentioned below). Thus ethylenediamine dichloride platinum(II), Pt(en)Cl_2,

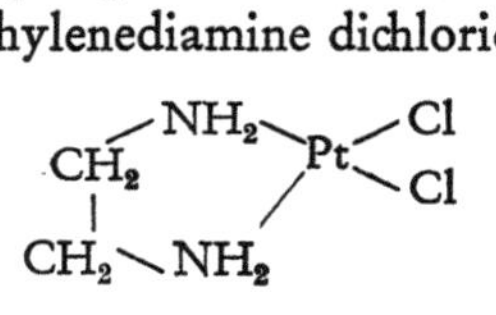

is also an effective anti-tumour compound [28]. In this case the thermodynamic stability of the amine groups is enhanced by the well known chelate effect [2].

Substitution reactions of planar platinum(II) compounds leading to the replacement of one of the ligands bound to the metal are known to follow a two path mechanism in aqueous solution, Fig. 2 ([2]). One path, denoted by the upper half of

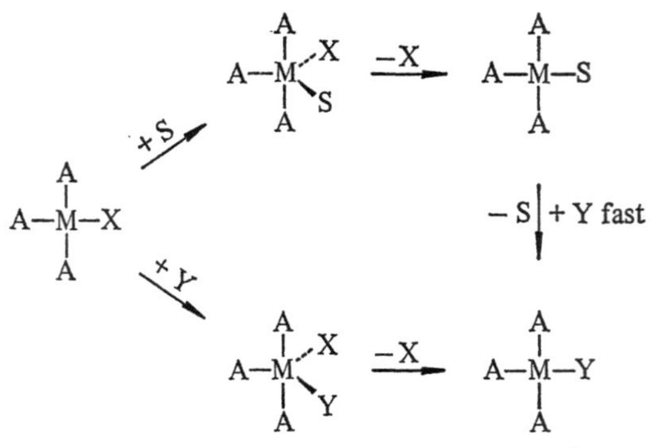

Fig. 2. The two path mechanism suggested for the reaction of a planar platinum(II) complex, MA_3X, with a nucleophile Y to give MA_3Y. The upper path involves the solvent [2]

Fig. 2, involves the replacement of a group bound to platinum, in this instance chloride ion, by the solvent, water, generating a solvated intermediate. The solvent molecule is subsequently eliminated by the incoming group, Y, which could in this case be one of the nucleophiles listed earlier. The alternative route involves the direct replacement of the leaving group by the attacking nucleophile, Y, without the participation of solvent. Both paths lead to total retention of the *cis* or *trans* starting configuration. The relative rates of each step determine the contribution of each path to the overall course of the reaction.

Table 1 a. Equilibrium constants for the replacement of one (K_1) and two (K_2) chloride ligands by water in *cis*- and *trans*-$Pt(II)(NH_3)_2Cl_2$

	K_1 (mole l^{-1})	K_2 (mole l^{-1})
cis	3.3×10^{-3}	4.0×10^{-5}
trans	3.2×10^{-4}	2.0×10^{-5}

From REISHUS, MARTIN [22].

Table 1 b. Relative number of moles of each of the species $Pt(NH_3)_2Cl_2$, $[Pt(NH_3)_2Cl(H_2O)]^+$, and $[Pt(NH_3)_2(H_2O)_2]^{++}$, in an aged solution of *cis*-$Pt(NH_3)_2Cl_2$ at 25° C under different conditions of chloride ion concentration

	$Pt(NH_3)_2Cl_2$	$[Pt(NH_3)_2Cl(H_2O)]^+$	$[Pt(NH_3)_2(H_2O)_2]^{++}$
Concentration of KCl = 0	2.7	4.4	1.0
Concentration of KCl = 1.0×10^{-3} (mole/l)	5.0	6.6	1.0

Initial concentration of *cis*-$Pt(NH_3)_2Cl_2$ = 10^3 mole/l. Ionic strength of Na_2SO_4 = 0.318.

From REISHUS, MARTIN [22].

The principles involved in such a substitution reaction are well illustrated by the reactions which both isomers undergo with the solvent, water, when in solution. The following two equilibria are set up with the equilibrium constants given in Table 1 a [1, 22].

$$Pt(NH_3)_2Cl_2 + H_2O \rightarrow [Pt(NH_3)_2Cl(H_2O)]^+ + Cl^-$$
$$[Pt(NH_3)_2Cl(H_2O)]^+ + H_2O \rightarrow [Pt(NH_3)_2(H_2O)_2]^{++} + Cl^-$$

Addition of chloride ion displaces these equilibria to the left. With the constants given in Table 1 a it is possible to calculate the relative number of moles of each of the three species present in solution at 25° C in the absence and presence of added chloride ion. The results are given in Table 1 b for the *cis* compound only at 25° C in Na_2SO_4 at an ionic strength of 0.318. These figures illustrate a distribution of species of different charge type whose concentration can be controlled with added chloride ion.

Radio-tracer experiments show no exchange of chloride ion with $Pt(NH_3)_2Cl_2$ although exchange does take place with $[Pt(NH_3)_2Cl(H_2O)]^+$ and $[Pt(NH_3)_2(H_2O)_2]^{++}$ suggesting that exchange occurs via the hydrolysed product, that is, via the upper pathway of Fig. 2 which involves the solvent. The replacement of chloride by water under these conditions is relatively slow taking place with a half-life of about 10 hours for both *cis* and *trans* isomers.

Aquo species, such as $[Pt(NH_3)_2(H_2O)_2]^{++}$ which can be formed rapidly from the dichloride by the addition of silver ions, undergo loss of protons to form the mono- and dihydroxy species, $[Pt(NH_3)_2(H_2O)(OH)]^+$ and $[Pt(NH_3)_2(OH)_2]^\circ$. The pK_a values for the *cis* and *trans* isomers, given in Table 1 c, enable calculation of the relative concentrations of these species at a given pH [7].

Table 1 c. pK_a values of *cis*- and *trans*-$[Pt(II)(NH_3)_2(H_2O)_2]^{++}$ at an ionic strength of 0.3 M $NaNO_3$

	pK_a^1	pK_a^2
cis	5.46	7.16
trans	4.32	7.00

From Grinberg *et al.* [7].

This information can be summarized in the following manner.

i) In water at pH 6.5 and zero chloride ion concentration we shall be observing the reactions of the species $[Pt(NH_3)_2(H_2O)(OH)]^+$.

ii) If the chloride ion resulting from the displacement reaction by water has not been removed from solution there will be a mixture of the species $Pt(NH_3)_2Cl_2$, $[Pt(NH_3)_2Cl(OH)]$, $[Pt(NH_3)_2(H_2O)(OH)]^+$ with roughly the concentrations given in Table 1 b.

In both instances (i) and (ii) direct replacement of OH^-, H_2O or Cl^- can occur by incoming groups which bind platinum strongly. Such groups may be sulphur or nitrogen, commonly.

iii) In a high chloride medium such as isotonic saline the species $Pt(NH_3)_2Cl_2$ will overwhelmingly predominate and replacement of chloride groups will be suppressed except for displacement by the strongest nucleophiles such as sulphur.

In none of this chemistry has a difference between the reactivity of the *cis*- and the *trans*-isomers been revealed. However, Fig. 1 shows clearly that the steric demands of the two isomers towards a *pair* of incoming groups which would replace chloride ions are very different. The leaving groups, the chloride ions of the *trans* isomer are on opposite sides of the molecule about 4 Å apart whereas in the *cis* isomer they are on the same side 3—3.5 Å apart.

b) Labilisation of Groups Trans to an Incoming Ligand. Although the amine groups bound to platinum(II) are usually kinetically inert they may become labile when opposite a ligand such as sulphur or cyanide. Since only the group *trans* to such a ligand is labilised the phenomenon affects the reactivity only of *cis* complexes. This difference between the reactivity of *cis*- and *trans*-compounds is the basis of a chemical test invented by Kurnakov to distinguish them. Here we choose one example only illustrating this type of reactivity as it has been seen by Russian workers in the reactions between *cis*- and *trans*-$Pt(NH_3)_2Cl_2$ and methionine.

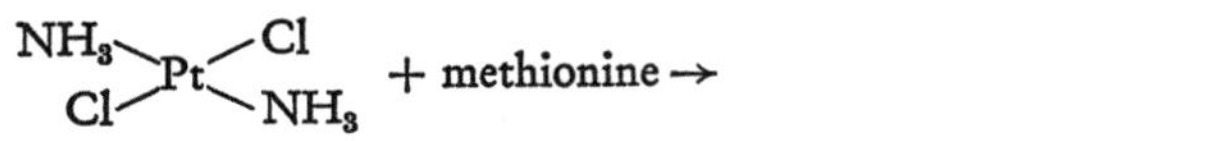

trans

Note that the amine groups are not displaced. (It is of interest that the methionine groups are retained even in high chloride ion concentrations. Only a stronger binding group such as iodide ion, I^-, will displace the methionine.)

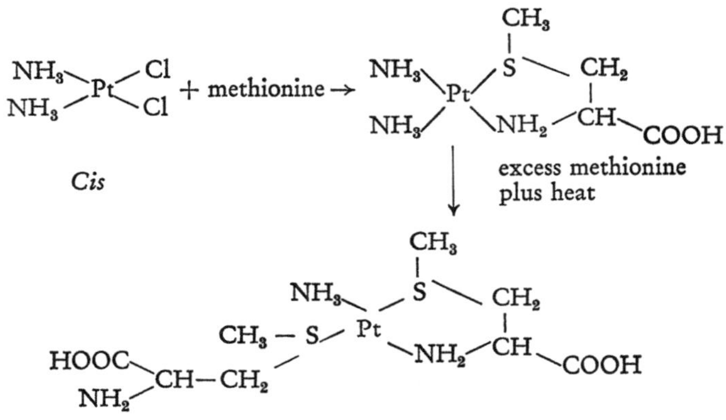

Cis

In this case a second methionine group can be added after an ammonia group has been labilised by the sulphur ligand *trans* to it. Note that the ammonia group *trans* to the amine of the methionine is not labilised [19].

A similar sequence of reactions has been observed between *cis*- and *trans*-$Pt(NH_3)_2Cl_2$ and 6-mercaptopurine, the strong *trans* labilising effect of the mercapto group stripping the ammonia ligands off the platinum in the case of *cis*-$Pt(NH_3)_2Cl_2$ only [8].

c) Reactivity of Groups Attached to Platinum. The reactivity of ammonia bound to platinum is not known to be high. However, metals such as Pt(II), especially Pt(IV), Hg(II), Au(III) and Ag(I) have strongly polarising properties that may activate attached groups. Thus ammonia when bound to Hg(II) will lose a proton at neutral pH although the pK_a of uncoordinated ammonia is greater than 30. Similarly the protons of the amine group in ethylenediamine are readily lost when the amine is co-ordinated to Au(III) or Pt(IV).

In a later section the chemistry of platinum "blue" compounds are discussed. Some of these are undoubtedly formed by the platinum catalysed addition of water to a co-ordinated acetonitrile ($CH_3—C≡N$) group.

d) Further modes of binding to planar platinum(II) complexes are possible. For example, planar platinum compounds have a weak affinity for groups in the fifth axial position. The tendency of planar platinum compounds to crystallise in columnar stacks arises from this tendency. However this type of binding is open to both *cis* and *trans* isomers alike. On the other hand the *cis* isomer has a considerably higher dipole moment than the *trans*. This may favour intercalation of *cis* compounds with DNA. At least one laboratory has looked for evidence of intercalation but none has so far been found.

Cis- and *Trans*-$Pt(IV)(NH_3)_2Cl_4$

a) Substitution Reactions. Nucleophilic substitution reactions of octahedral platinum(IV) compounds are much less well studied than those of platinum(II) compounds. Their reactions are catalysed by traces of Pt(II) ion or by light and hence the mechanisms of their reactions are difficult to study [2]. However, it is known that both isomers hydrolyse in aqueous solution, the *cis* isomer reacting more rapidly than the *trans* [21]. Early work reported in Gmelins Handbuch [10] suggests an intriguing difference between the reactivities of the ammonia groups of the *cis* and *trans* isomers that should be re-investigated in view of the drug action of the *cis* isomer. The *trans* isomer reacts with pyridine in the warm losing some of its chloride ions to form $[Pt(IV)(NH_3)_2(pyridine)_2Cl_2]Cl_2$. The stereochemistry of this species is not reported. Excess ammonia will also replace two chloride ions of the *trans* isomer to yield the species $[Pt(IV)(NH_3)_4Cl_2]Cl_2$. Again no stereochemical information is given. If these reactions are shown to involve the replacement of a pair of *trans* chloride ions the lack of anti-tumour activity of the *trans* isomer would find a ready explanation. In contrast to these reactions, the *cis* isomer reacts with warm pyridine with the loss of its ammonia groups to give $Pt(IV)(pyridine)_2Cl_4$. This lability of ammonia groups is surprising and is worthy of further study since this reaction, the loss of two *cis* groups, has obvious parallels with the *cis*-platinum(II) complexes except that the ammonia ligands are playing the role of leaving groups.

b) Redox Reactions. Both *cis*- and *trans*-$Pt(IV)(NH_3)_2Cl_4$ can be reduced to the isomers of $Pt(II)(NH_3)_2Cl_2$ with retention of configuration at a potential close to zero volts. It is therefore quite possible that the active forms in a biological system of the platinum(IV) complexes are the platinum(II) analogues.

From this review of the chemistry of biologically active and inactive platinum compounds we conclude that both *cis*- and *trans*-$Pt(NH_3)_2Cl_2$ and many of the analogues reported in the earlier papers exchange two of their ligands (chloride ion most often) readily for groups such as $R{-}S{-}CH_3$, RS^-, $\rangle N$ and $R{-}NH_2$ to form links which can be very stable. Although both isomers can thus act as bi-functional agents they have very different stereochemical demands as readily appreciated from Fig. 1, which could account for their very different drug action. The role of platinum(IV) compounds as drugs does not yet emerge clearly from their chemistry. Some basic biochemical questions concerning the nature, and especially the oxidation state, of the metal ion *in vivo* must first be answered.

Reaction of Platinum Drugs with Ligands Present in a Biological Medium

Early radiotracer studies on the distribution of platinum complexes in whole bacteria grown in a medium containing the metal ion revealed the following facts.

The bacteriocidal complex, $[Pt(IV)Cl_6]^{2-}$, is taken up almost entirely by the cytoplasmic protein whereas the filamentous forming species, $[Pt(IV)(NH_3)_2Cl_4]$ of unknown ratio of *cis* to *trans* isomers, was bound to the nucleic acid fraction and to metabolic intermediates which comprise amino-acids, peptides and polynucleotides. Lipids contained a surprisingly small fraction of any of the species [23].

In the following sections we discuss the interaction of platinum(II) compounds only with each of the components, nucleic acids and protein, in turn.

Binding to Nucleic Acids

The observation of appreciable binding of platinum compounds *in vivo* to nucleic acids has since been given added significance by several experiments. Thus [24] showed that *cis*-$Pt(NH_3)_2Cl_2$ induced prophage from lysogenic bacteria while [11]

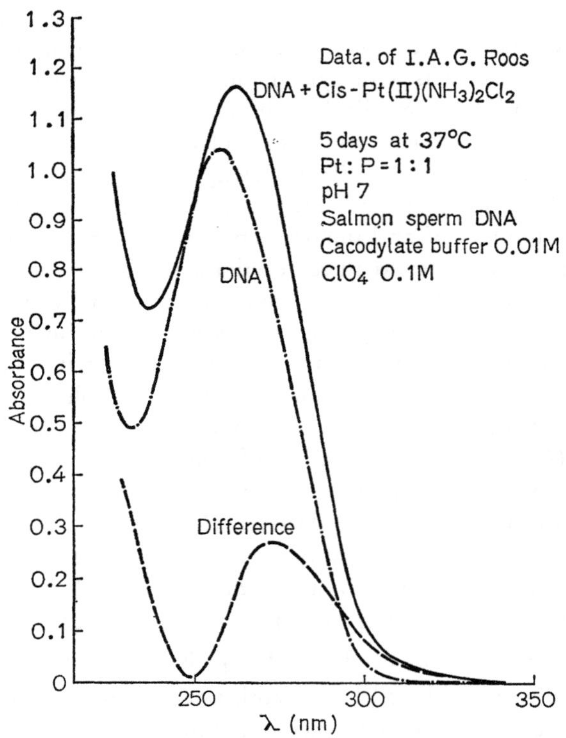

Fig. 3. The absorption spectrum (1 cm path) of salmon sperm DNA in the presence and absence of *cis*-$Pt(II)(NH_3)_2Cl_2$. Reaction was allowed to proceed for 5 days at 37° C at pH 7 in cacodylate buffer (0.01 M) and $NaClO_4$ (0.1 M). The Pt : P ratio is unity [30]

were able to demonstrate that DNA synthesis in human amnion cells, AV_3, was selectively inhibited by *cis*-$Pt(NH_3)_2Cl_2$ but only by much higher concentrations of *trans* isomer. It is thus of great interest to uncover the modes of interaction of platinum compounds with DNA. A number of studies have now been carried out with this aim in view. However, Russian workers had already demonstrated that $[Pt(II)Cl_4]^{2-}$

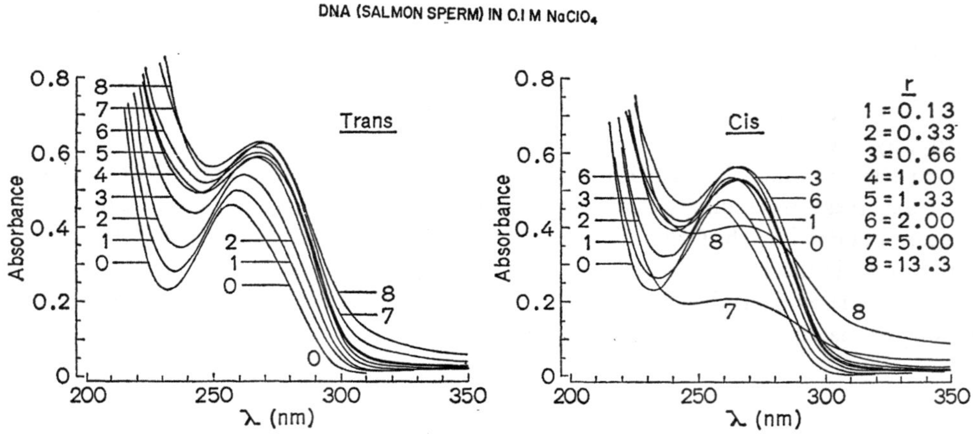

Fig. 4. The absorption spectra (1 cm path) of solutions of salmon sperm DNA in 0.1 M $NaClO_4$ in the presence of *trans*- and *cis*-$Pt(II)(NH_3)_2Cl_2$ after equilibrium had been achieved. r = Pt : P ratio [18]

reacts with the bases of DNA and that the rate of reaction increases with the G + C content of the DNA. As they were also able to show that apyrimidinic acid, adenine and hypoxanthine react with $[Pt(II)Cl_4]^{2-}$ they concluded that purines were important in the binding [34, 20].

The simplest method to demonstrate that both *cis*- and *trans*-$Pt(NH_3)_2Cl_2$ interact with DNA is spectrophotometrically. Fig. 3 shows the spectrum of salmon sperm DNA after reaction with *cis*-$[Pt(NH_3)_2Cl_2]$ in 0.1 M $NaClO_4$ and 0.01 M cacodylate buffer at pH 7 at a ratio of Pt : P of 1. Under these conditions the reaction is slow, taking about five days for equilibrium to be reached. The red shift of the absorption spectrum of DNA in the presence of *cis*-$Pt(NH_3)_2Cl_2$ is quite unambiguous proof that the aromatic bases of DNA are being attacked since the absorption spectrum of DNA at this wavelength arises from the aromatic bases. Conversely no information is obtained about possible binding of platinum to sugar or phosphate residues.

If this reaction is attempted in 0.1 M NaCl no change in the absorption spectrum is obtained over a period of at least five days. However, if after the reaction has been allowed to proceed in the absence of added chloride ion, the chloride ion concentration is subsequently raised to 0.1 M no reversal of the binding can be detected spectrophotometrically. These results are readily interpreted in terms of our earlier discussion of the chemistry of platinum. In the presence of a high concentration of chloride ion nucleophilic displacement of chloride bound to platinum by groups such as $>$N, —NH_2 or $>$C=O present on the purine and pyrimidine bases does not take place to a significant extent. However, once binding of such groups has been permitted to occur by lowering the chloride ion concentration then the resultant link cannot be displaced by a subsequent increase in chloride ion concentration because of its high thermodynamic and, most importantly, its kinetic stability.

Such experiments tell us therefore that *cis*-$Pt(NH_3)_2Cl_2$ has reacted with the bases of DNA with the displacement of at least one chloride ion. Both *cis*- and *trans*-$Pt(NH_3)_2Cl_2$ react in this manner as can be seen from Fig. 4. At high ratios of Pt : P with the *cis* isomer DNA is precipitated from solution leading to the observed loss of optical density [18]. A similar study has been published by HORACEK and DROBNIK [13] with only *cis*-$Pt(NH_3)_2Cl_2$. They were able to demonstrate the lowering of the affinity of DNA for the platinum compound by chloride ion and, also most significantly, by phosphate. Nitrate and sulphate showed little effect. The latter two anions are expected to bind platinum(II) rather poorly but evidence is emerging (see earlier reports in these proceedings) that phosphate is able to bind platinum. Good evidence was also obtained by the Czechoslovakian workers for the reaction of *cis*-$Pt(NH_3)_2Cl_2$ with DNA via the hydrolysed species, that is, the upper path of Fig. 2.

So far the evidence has not revealed a clear cut difference between the modes of binding of the *cis* and *trans* isomers to DNA.

Binding to Purines and Pyrimidines*

The next stage is clearly to enquire which bases are vulnerable to attack by platinum compounds, what sites are occupied, and whether a difference between the binding of the two isomers can be demonstrated at this level. Two studies designed

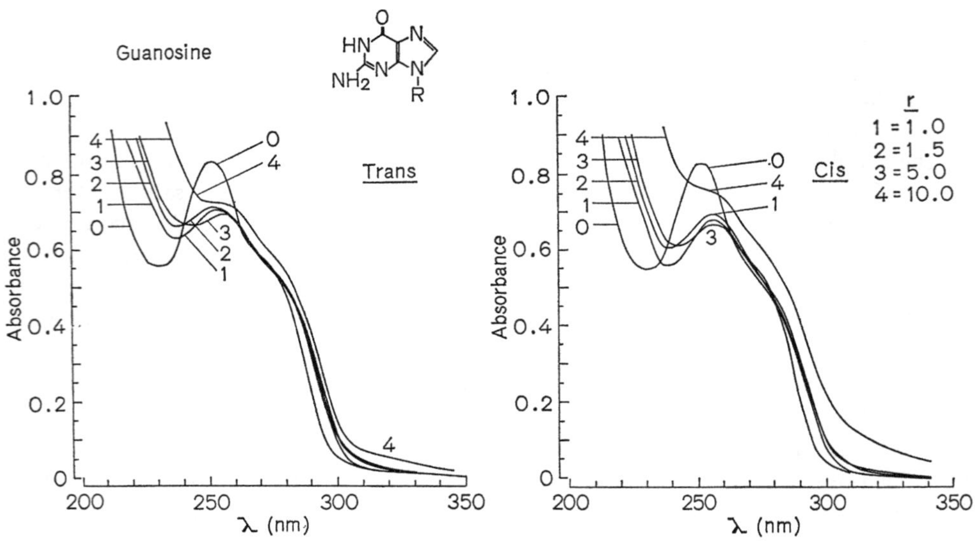

Fig. 5. The absorption spectra (1 cm path) of solutions of guanosine (10^{-5} M) and *trans*- and *cis*-$Pt(II)(NH_3)_2Cl_2$ after equilibration at 37° C in 0.1 M $NaClO_4$. r = Pt : P ratio [18]

to provide answers to these questions have been undertaken. The first, by MANSY *et al.* [18] was a spectrophotometric study of the binding of *cis*- and *trans*-$Pt(II)(NH_3)_2Cl_2$ to nucleosides and has provided information about the sites of platinum binding at the bases under equilibrium conditions. A second investigation, by ROBINS [25] looked at the kinetics of reaction between nucleosides and the radioactively labelled

* Nomenclature. The convention followed here for the numbering of the pyrimidine ring is that used by*Chemical Abstracts*. Under this system uracil is 2,4-dihydroxypyrimidine.

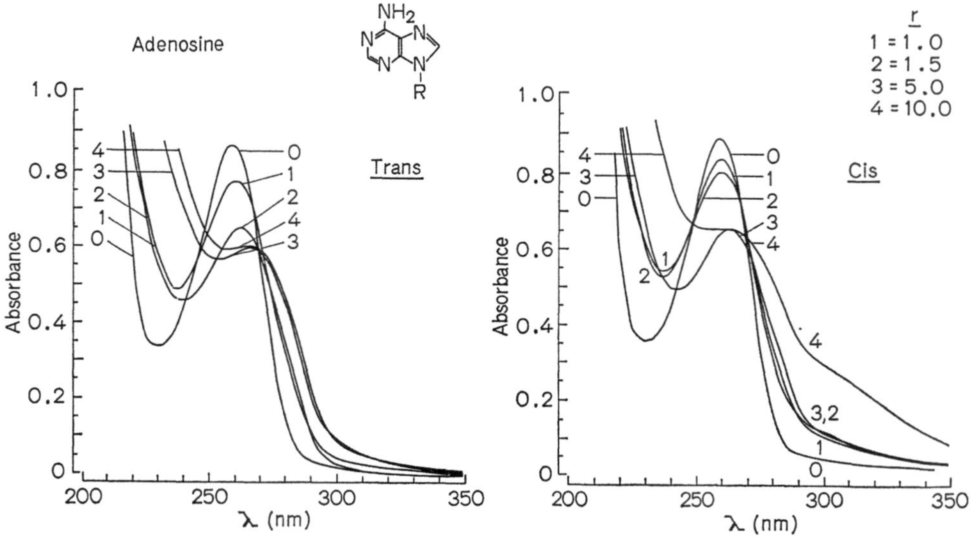

Fig. 6. The absorption spectra (1 cm path) of adenosine (10^{-5} M) equilibrated at 37° C with *trans*- and *cis*-Pt(II)$(NH_3)_2Cl_2$. r = Pt : P ratio [18]

anti-tumour compound Pt(II)(^{14}C-en)Cl_2 using a chromatographic technique to separate the products.

We consider first the attempts to locate the binding sites on the purine and pyrimidine rings. In this case nucleosides were employed to ensure that the N-1 and N-9 positions of the pyrimidine and purine rings, respectively, were blocked to attack by metal ion. These are sites with a high affinity for many metal ions (and protons) which are unavailable in nucleic acids. Both isomers of $Pt(NH_3)_2Cl_2$ were investigated in a search for differences between their reactions with bases.

Both guanosine and adenosine react with the *cis* and *trans* isomers, Figs. 5 and 6. At high ratios of Pt to base some background absorption from the unreacted platinum compound is evident. It is striking that whereas both isomers give products with virtually identical spectra on reaction with guanosine a distinct difference is apparent between the products derived from adenosine. The long wavelength absorption at ~310 nm arising from the *cis* isomer and adenosine is noteworthy. A similar red-shifted band has previously been seen in complexes formed between Ag(I) and Hg(II) and adenosine. Although the origin of the band is not clear most studies agree that it is symptomatic of binding to the 6-amino group of adenosine [9, 31].

In order to define further the sites of binding on the purine rings two methods were used by Mansy *et al.* [18]. The purine ring system contains a number of potential binding sites for metals. Thus in adenosine any one of four nitrogen atoms are possible ligands for a metal ion. However it is possible to block these sites to reaction with metal ions either by methylation or by protonation of the nitrogen atoms. In the former case some ambiguities may arise since methylation of the ring may alter the affinity of the remaining open sites towards metal ions. However, to guard against this fact giving misleading results a check was obtained by blocking sites sequentially with protons. This method assumes a knowledge of the sites of protonation of the

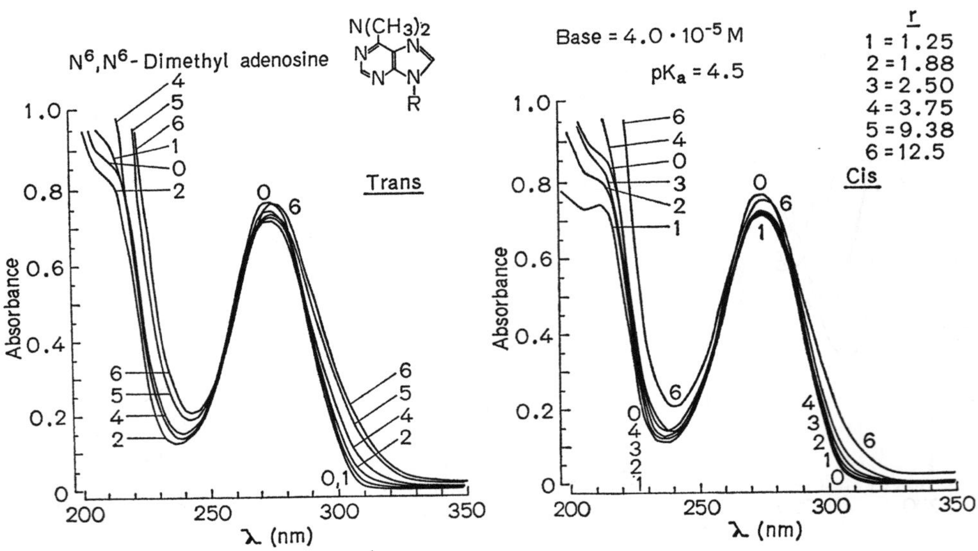

Fig. 7. The absorption spectra (1 cm path) of N[6], N[6]-dimethyladenosine (4×10^{-5} M) equilibrated with *trans*- and *cis*-Pt(II)$(NH_3)_2Cl_2$. r = Pt : P ratio. (Taken from same reference as Fig. 6)

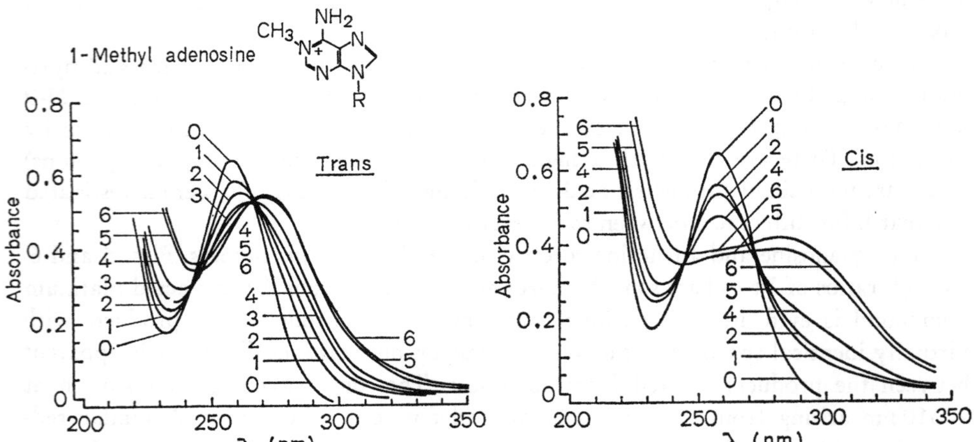

Fig. 8. As for Fig. 7 except that 1-methyladenosine is used instead of N[6], N[6]-dimethyladenosine

ring nitrogen atoms under given pH conditions. These facts are now well established. Here we relate in detail only the results obtained using methylated bases since the pH dependent studies broadly confirm the results.

Consider first the reactions of *cis*- and *trans*-Pt$(NH_3)_2Cl_2$ with methylated adenosines. Little reaction is apparent with N[6], N[6] dimethyladenosine and either isomer, Fig. 7. This result eliminates N 3 as a site of attack and strongly implicates the 6-amino group. The long wavelength absorption found with adenosine and the *cis* compound is notably absent. On blocking position 1 in adenosine, Fig. 8, extensive reaction with both isomers proceeds and with the *cis* isomer a species having a strongly red shifted absorption band is seen. Since the *cis* isomer only is able to form a bi-dentate chelate we suggest the products are complexes of the following structure:

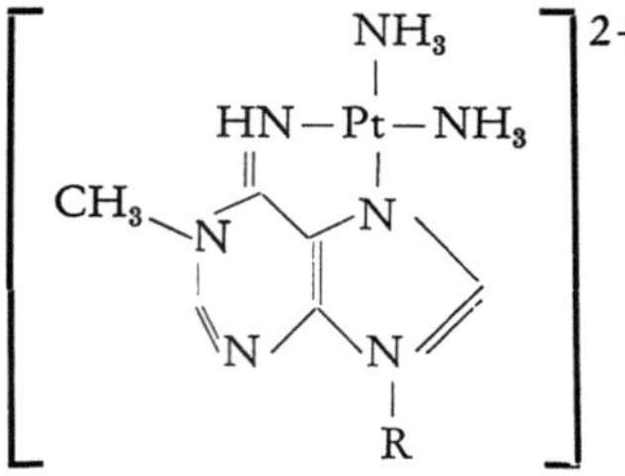

cis

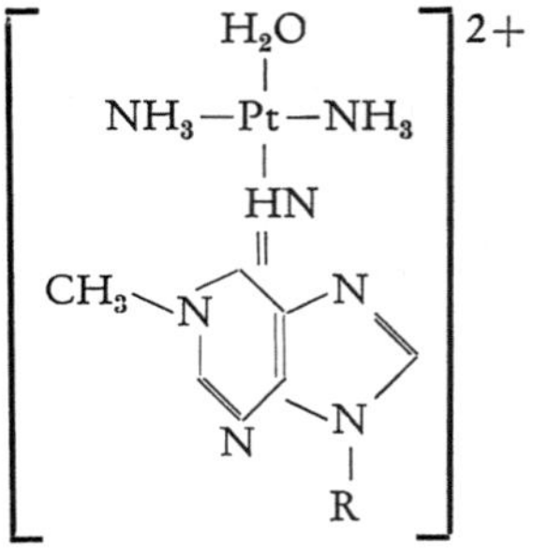

trans

In a study of the spectra as a function of the mole ratios of metal to base good isosbestic points were obtained in the case of 1-methyladenosine suggesting that a 1 : 1 complex was being formed.

We note that the pK_a of the 6-amino group of 1-methyladenosine is 7.6 whereas this pK_a has not been measured in adenosine itself since it is undoubtedly too high. Some estimates have placed it above 16. This dramatic lowering of the pK_a arises because of the positive charge placed on the ring by the methyl group. This may well account for the overwhelming binding to the 6-amino group in this case.

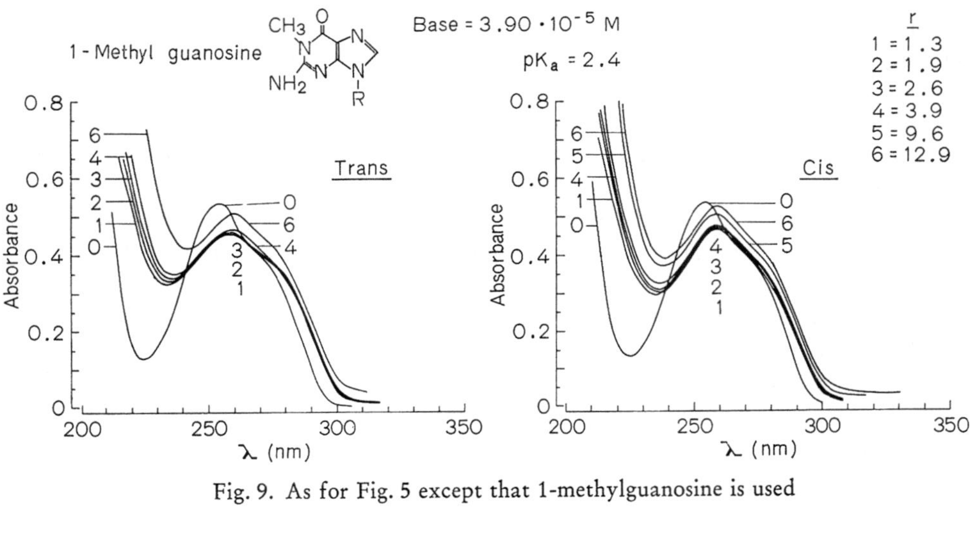

Fig. 9. As for Fig. 5 except that 1-methylguanosine is used

pH dependence studies of the binding of the isomers to adenosine itself give strong support to the modes of binding suggested by the studies with methylated bases. That is, the *cis* isomer can form a bi-dentate chelate between 6-NH_2 and N-7 positions. Additional evidence is obtained that N-1 is a site of high affinity for both isomers when open.

In contrast to these results, in the reactions of the *cis* and *trans* isomers with guanosine and methylated guanosine no difference is apparent between the spectra of

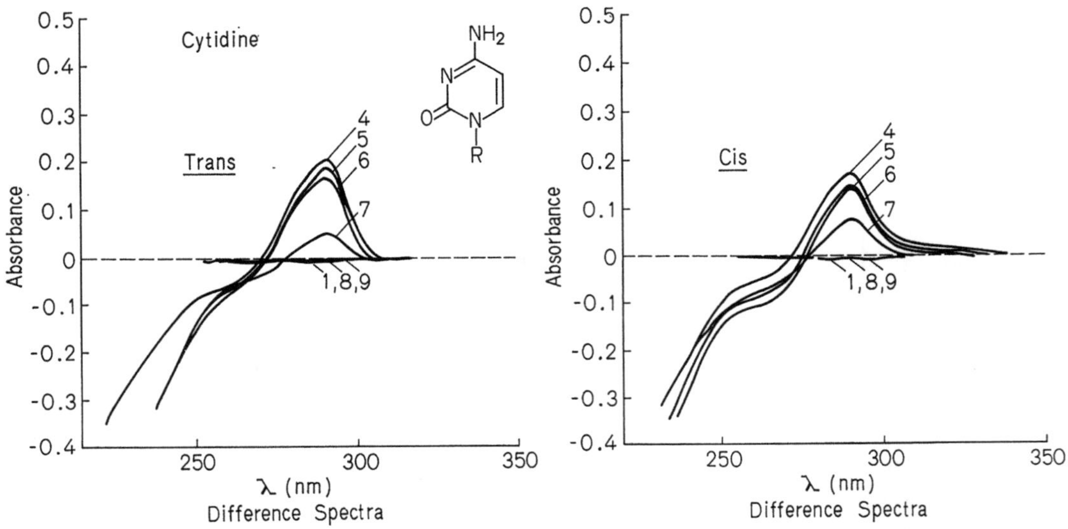

Fig. 10. The difference spectra of cytidine after equilibration with *trans*- and *cis*-Pt(II)-$(NH_3)_2Cl_2$. Samples were contained in 1 cm path with equal concentrations (10^{-5} M) of cytidine and platinum in sample and reference path. The numbers on the spectra indicate the pH values of the solutions [18]

the products. Blocking of the 1 position as in 1-methylguanosine has little effect on the binding of either isomer, Fig. 9. Thus the N-1 position can be eliminated from consideration, a not surprising result since this site is blocked with a proton at neutral pH (pK_a(N-1) = 9.7). No reaction is observed spectrophotometrically between either isomer of platinum and 7-methylguanosine. This result should be compared with that reported by Robins [25] who detected a binding between 7-methylguanosine and $Pt(^{14}C\text{-en})Cl_2$ radiochemically. Possibly this binding is to the ribose group. However the evidence for the involvement of the N-7 position of guanosine is quite clear. The similarity of the spectra of the products from guanosine and both isomers are an indication that no form of bidentate chelation of the *cis* isomer between N-7 and the 6-keto groups is occurring as might have been expected. The pH dependence studies reaffirmed this conclusion. However, recent mass spectrometric studies by Thomson and Roos [30] give some indication of such a species being formed by the *cis* isomer. This is an important point to clarify in view of the kinetic studies of Robins [25] which suggest that guanosine is the base most rapidly bound by platinum in DNA.

Inosine reacts in a similar manner to guanosine with both isomers of platinum, a result which suggests that the 2-amino group of guanosine is an unimportant binding site of platinum.

Finally we consider binding of platinum to the pyrimidine ring. Cytidine reacts with both *cis*- and *trans*-$Pt(NH_3)_2Cl_2$, Fig. 10, whereas 3-methylcytidine does not. This result shows that the N-3 position is attacked, a view which is supported by pH dependence studies. Again the similarity of the spectra obtained with both isomers argues against there being any bi-dentate chelation by the *cis* isomer. These studies have revealed no binding to the 4-amino group or the 2-keto group of cytidine. However there is evidence that the 4-amino group may become bound when the cytidine ring is part of a dinucleotide (see later).

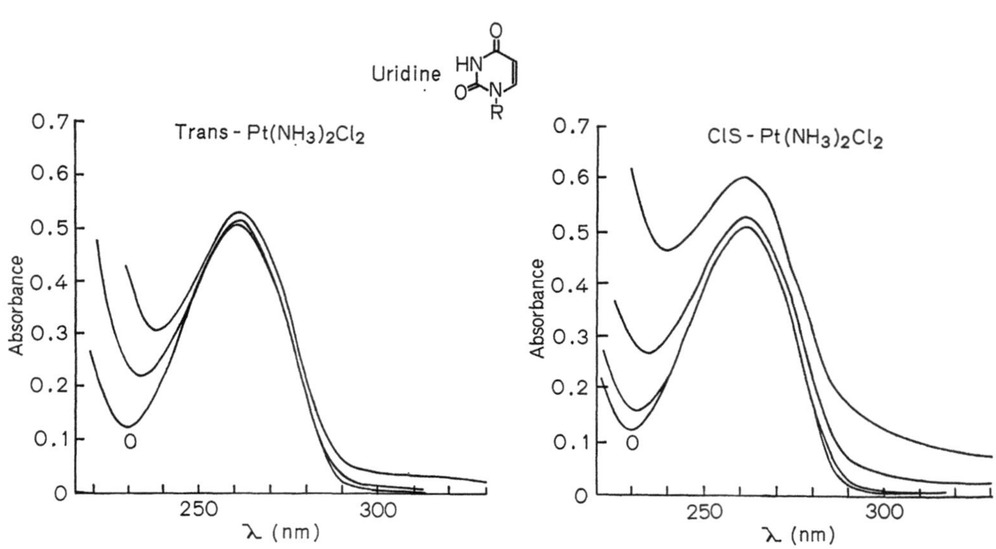

Fig. 11. The absorption spectra of uridine (10^{-4} M) after 5 days incubation at 37° C in 0.1 M $NaClO_4$ with *trans*- and *cis*-Pt(II)$(NH_3)_2Cl_2$. 1 cm cell. The r values are the ratio of Pt : P and are the same as in figure 5. The background contribution from unreacted *cis* isomers is seen in curve 4 [18]

In a spectrophotometric study of the interaction of uridine with *cis*- and *trans*-Pt$(NH_3)_2Cl_2$ no reaction was at first found, Fig. 11. But in a subsequent study of the interaction of the *cis* isomer with polyuridine, Mansy and Rosenberg [16] observed that a blue-green colour was slowly formed over the course of three weeks. Later they were able to observe this colour after reaction of *cis*-[Pt(II)$(NH_3)_2(H_2O)_2]^{2+}$ with uracil and many of its analogues. In a mass spectrometric investigation, Roos and Thomson [30] searched for interaction between 1-methylthymine (a volatile nucleoside analogue) and several effective drug analogues of the type *cis*-[Pt(II)(R—$NH_2)_2Cl_2$]. No interaction was at first detectable but on leaving an aqueous solution of 1-methylthymine and bis-cyclopropylamine Pt(II) dichloride to stand in the dark, a purple colour slowly formed in the solution.

These observations have given rise to a number of investigations of the nature of these blue materials stimulated on the one hand by the observations of Rosenberg and Van Camp [28] that some of the blue compounds have appreciable anti-tumour activity and on the other by the need to have a nucleotide specific label of DNA which is coloured and hence readily observed. A short review of the chemistry of "blue" platinum complexes is given at the end of this article.

The lack of a reaction initially between uridine or thymidine and the platinum(II) isomers most likely arises because the N-3 position is blocked by a strongly bound proton ($pK_a = 9.8$ for thymidine). The oxygen atoms in the keto form are not expected to have a high affinity for platinum. These expectations are well borne out by the work of the Russians on the interaction between the antimetabolite, 5-fluorouracil (FU), and *cis*- and *trans*-Pt$(NH_3)_2Cl_2$. Reaction takes place only at high pH (0.14 N LiOH) after one of the protons of FU has been removed. The products are *cis*- and *trans*-Pt$(FU)_2(NH_3)_2$, *white* flocculent precipitates. The uracil groups are so strongly bound that boiling alkaline iodide will not remove the FU ligands. However, boiling HCl does displace FU to regenerate the starting material in both cases, showing that the amine groups had not been displaced. It is suggested, but not proven, that the FU is bound via a Pt-N bond. It is certain however that the FU is bound in a monodentate manner only since two FU molecules are bound for the loss of two chloride ions [5].

Returning now to the kinetic studies of Robins [25, 26] which showed that the reactions of all the bases were second-order and led to the formation of a complex with a Pt : base ratio close to unity. An exception was the reaction with adenosine

Table 2. Kinetic data for the reactions of Pt(^{14}C-en)Cl_2 with nucleic acids and their components (from Robins [26])

Constituent or analogue	Second order rate constant $l \cdot M^{-1} \cdot sec^{-1} \times 10^{-3}$	Half-time of reaction (hours)
Guanosine	106	6.7
Deoxyguanosine	62	12
7-methylguanosine	24	30
7-methylguanine	100 (initial)	ca. 7
Guanine	10	70
Adenosine	6	120
Adenine	8	90
Adenosine 5′-monophosphate	330	2
Adenosine triphosphate		70—90 (Two species 250 detected)
$A_{3'}p_{5'}A$	122	
Poly-A	410	
Poly-d(AT)	80	
Polyguanylic acid	150	
Poly dG—poly dC	70	
DNA		
0—24 hours	100	
1—6 hours	6	

Estimated for rate of disappearance of Pt(en)Cl_2 at 37° C.

triphosphate (ATP) where an additional product with a Pt : base ratio of 2 was found. The results are summarized in Table 2. The rates vary over a wide range, those with guanosine and its derivatives are the most rapid with the notable exception of adenosine 5′-monophosphate which reacts with the shortest half-life found, 2 hours.

Although reaction is taking place with the ring nitrogen atoms of the bases the rates are seen to be very sensitive to the presence of additional groups attached to the ring. Thus the ribose and deoxyribose groups increase the rate of reaction with guanosine almost 10 fold. It is not known whether this is a steric effect or an electronic one arising from the presence of the alkyl group at N-9. Since the site of metallation is N-7 it is possible that the effect is electronic.

The rates of reaction with the adenine ring do not show enhancement due to the presence of sugar residues possibly because metallation is occurring in the other ring at N-1 and between 6-NH_2 and N-7. However, the effect of a monophosphate group in enhancing the rate is of great interest. It may be related to the recent observation that the manganese ion, Mn^{2+}, can simultaneously bind the phosphate group and the N-7 position of adenosine 5′-monophosphate. Other metal ions, Co^{2+}, Ni^{2+} are also thought to bind in this way [15].

On addition of further phosphate groups, as in ATP, a slowing down of the rate of reaction was found by Robins [25]. This could well arise from the preferential interaction of the platinum *cis* isomer with the triphosphate group in a manner similar to that observed with pyrophosphate (reported earlier in these proceedings). This additional mode of binding would then account satisfactorily for the second product found in the reaction with ATP.

In conclusion then, studies with *cis*- and *trans*-$Pt(NH_3)_2Cl_2$ have revealed a number of possible sites of binding by platinum on the bases of nucleic acids but only in one case, that of adenosine, is a clear difference revealed between the modes of binding of the two isomers. The rate studies have further shown that the speed of reactions with a particular base can be markedly dependent upon the surrounding group which may be able to participate in the reaction path.

Binding to Oligonucleotides

When the purine and pyrimidine bases are built into the helical structure of a nucleic acid it is possible that sites reactive towards platinum will be suitably aligned to form two links to one platinum ion. Since the stereochemical demands of the two isomers are very different the same site is unlikely to accommodate both isomers. In order to investigate the likelihood of forming a link between two adjacent bases with a platinum ion, Roos *et al.* [29] investigated the binding of *cis*- and *trans*-$Pt(II)(NH_3)_2Cl_2$ to some dinucleotides.

Various physical techniques [14, 35] including recently X-ray crystallography [4] have shown that dinucleotides can exist in two forms, a stacked form, with the planes of both bases parallel and vertically above one another, and an unstacked conformation in which the bases are swung away from each other. The equilibrium between these two conformers is often steeply temperature dependent, the stacked form predominating at low temperature, and also pH dependent so that after protonation of the bases electrostatic repulsion unstacks the dimer. A dinucleotide in the stacked conformer places a steric restraint upon the groups which can be bound simultaneously to one platinum atom. Fig. 12 gives an example of a possible stacked conformation of the dinucleotide adenosine 3′-phosphate-5′-adenosine, $A_{3'p5'}A$. It is of interest to detect just such modes of binding since in the organised structure of a helical nucleic acid many of the binding sites freely available in nucleosides will become sterically hindered and sites less favoured in nucleosides may become significant. A technique is needed which will detect these special modes of binding. In the case of the dinucleotides the technique must pick out the binding in the presence of the binding of platinum to a single adenosine ring, for example. The circular dichroism spectrum (c.d.) of the stacked form of the dinucleotide, ApA, is a highly characteristic biphasic signal whereas the unstacked conformer has a weak monophasic signal. If a

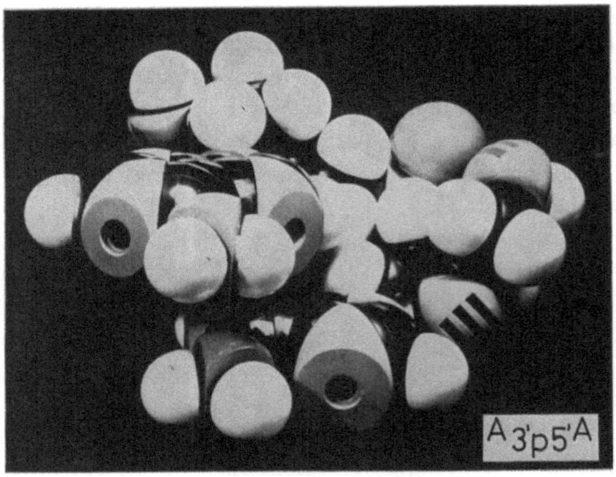

Fig. 12. A Corey-Pauling-Koltun model of the stacked conformation of adenyl-3'-ribo-phosphate-ribo-5'-adenosine ($A_{3'p5'}A$). The 6-amino groups on each adenine ring are facing the reader

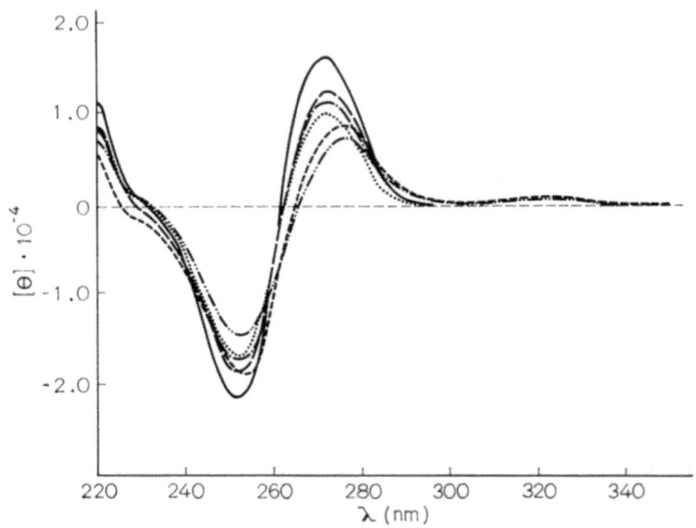

Fig. 13. The circular dichroism spectrum of $A_{3'p5'}A$. Pt : P = 1 : 1. Dinucleotide concentration 4×10^{-5} mole dm^{-3}; temperature 37° C. $[\Theta]$ = molar ellipticity. $A_{3'p5'}A$ (———), plus *cis*-$Pt(NH_3)_2Cl_2$ (— — —), plus *trans*-$Pt(NH_3)_2Cl_2$ (·······), plus *cis*-$Pt(enim)_2Cl_2$ (—··—··—), plus *cis*-$Pt(NH_3)_2Cl_4$ (— — —), plus *trans*-$Pt(NH_3)_2Cl_4$ (—·—·—). (enim = ethyleneimine)

single platinum complex reacts with two bases of a single dinucleotide in the stacked conformation a temperature independent biphasic c.d. signal will be observed (provided that the platinum ion is not displaced at high temperatures). The effect of the platinum ion will be to clip the dimers into a conformationally rigid stacked form. On the other hand if a platinum complex reacts with only one base unstacking of the dimer will result since a positive charge will be placed on the ring.

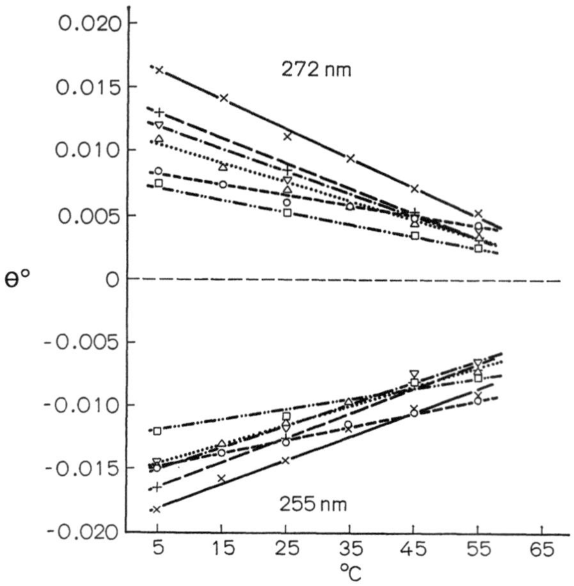

Fig. 14. Temperature dependence of the circular dichroism of $A_{3'p5'}A$ with platinum complexes. $A_{3'p5'}A$ (———), plus *cis*-$Pt(NH_3)_2Cl_2$ (— —○— —), plus *trans*-$Pt(NH_3)_2Cl_2$ (· · · △ · · ·), plus *cis*-$Pt(enim)_2Cl_2$ (— · · □— · ·), plus *cis*-$Pt(NH_3)_2Cl_4$ (— + —), plus *trans*-$Pt(NH_3)_2Cl_4$ (— ·▽— ·). Conditions the same as for Fig. 13

Fig. 13 shows the c.d. spectra of the dinucleotide $A_{3'p5'}A$ after reaction with a number of platinum drug analogues, both Pt(II) and Pt(IV) *cis* and *trans* isomers. The temperature dependence of the rotational strength at the peak and trough of each spectrum is plotted out in Fig. 14. Both *cis* isomers tested give appreciably reduced slopes compared with unreacted dinucleotide and *trans* isomers. This points to the presence of a temperature independent stacked dinucleotide in the case of the *cis* isomers. However, in order to show quite unambiguously that the temperature independent form reducing the slope is stacked the pH was lowered to 1. The adenine ring becomes protonated, on N-3, at this pH so that $A_{3'p5'}A$ unstacks, and the biphasic c.d. signal is abolished. Fig. 15 shows that in the presence of *cis* isomers of platinum only, the biphasic c.d. spectrum is retained at pH 1. Thus we conclude that an acid stable interbase link is present, preventing the bases unstacking. In this way it was possible to show that an interbase link can also be introduced into the dimer $A_{3'p5'}C$ by *cis* platinum drugs. However, there is an important difference between the pH stability of the links in ApA and ApC that gives useful further information about the sites of binding on the rings which are linking the bases. The interbase link of ApA is alkali stable, at pH>10, whereas the link is removed from ApC at this pH.

Inspection of models, Figs. 12 and 16, shows that the stereochemically possible sites for the attachment of a *cis* platinum compound to insert an interbase crosslink are, in the case of $A_{3'p5'}A$, between (i) the 6-NH_2 groups on both bases (ii) N-1 (3') and 6-NH_2 (5') or (iii) 6-NH_2 (3') and N-7 (5') and, in the case of $A_{3'p5'}C$, between (i) N-1 (3') and 4-NH_2 (5') and (ii) 6-NH_2 (3') and 4-NH_2 (5').

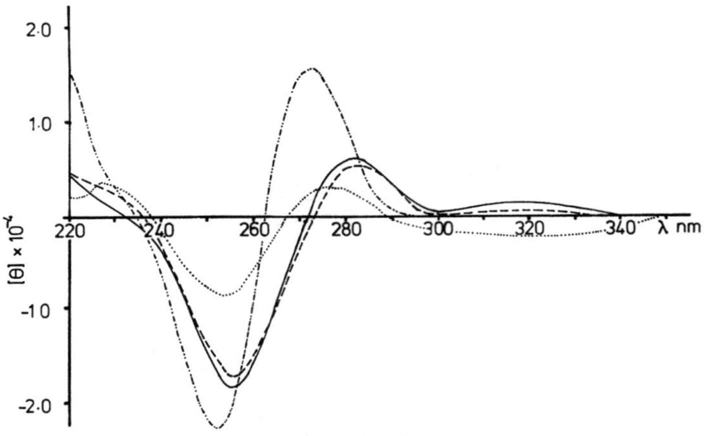

Fig. 15. pH dependence of the circular dichroism spectrum of $A_{3'p5'}A$ with *cis*-Pt(enim)$_2$Cl$_2$, Pt : P, 1 : 1. Dinucleotide concentration 1×10^{-4} mole dm^{-3}; temperature 37° C. pH 6—7 (———), pH 11—12 (— — —), pH 1 (········), $A_{3'p5'}A$ at pH 6—7 (—··—··—)

At high pH the reacting platinum species will be predominantly [(amine)$_2$Pt(II)·(OH)$_2$] which on linking two amino groups of a dinucleotide will yield the product

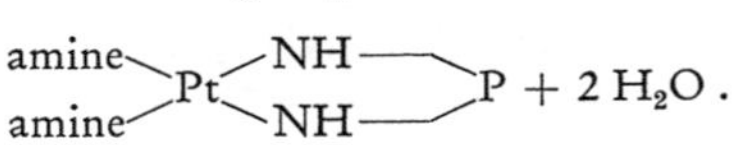

This reaction is pH independent at high pH. On the other hand the following reaction is reversed at high pH.

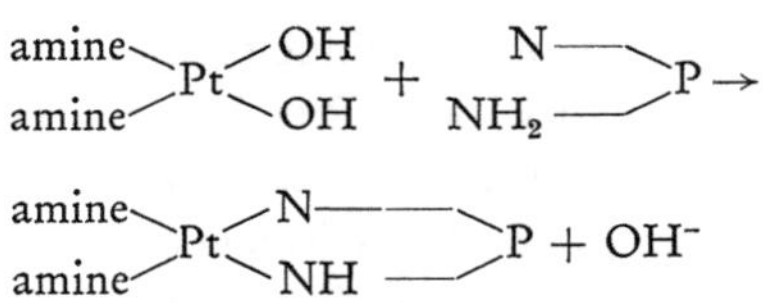

The crosslink in $A_{3'p5'}A$ is maintained at high pH. The only one of the three stereochemically feasible links listed above which will remain intact is (i) that between the 6-NH$_2$ groups on each base. In $A_{3'p5'}C$ the binding is alkali sensitive and hence the link must be between the 4-NH$_2$ group of the cytidine ring and the N-1 position of the adenine ring.

Thus we have been able to demonstrate that the organisation of binding groups into a suitable conformation provides sites for the bi-dentate chelation of *cis*-Pt(NH$_3$)$_2$Cl$_2$ which are not occupied by the *trans* isomer. There are undoubtedly many such sites suitably disposed in DNA itself. One site that can be seen in models of DNA is of particular interest in view of the cross-linking of two 6-amino groups of adenine rings demonstrated here. Fig. 17 shows the orientation of the 6-amino groups on two adenine rings in a portion of DNA with the sequence A, T in the 3′-5′ direction. The amino groups are vertically above one another with the interplanar base separation of ~3.5 Å, which is the distance required to bridge the two *cis* ligand positions on square planar platinum(II) compounds (Fig. 1). Thus a platinum link between the two amino groups would constitute a very stable inter-strand crosslink.

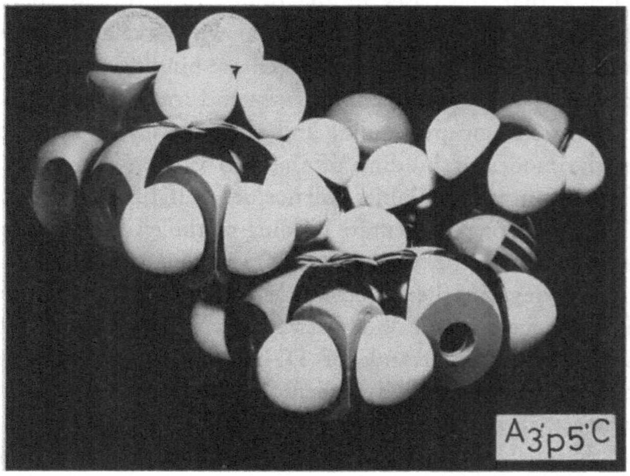

Fig. 16. A Corey-Pauling-Koltun model of adenyl-3'-ribo-phosphate-ribo-5'-cytidine in the stacked conformation. The cytidine ring is uppermost with the amino groups of both rings facing the viewer

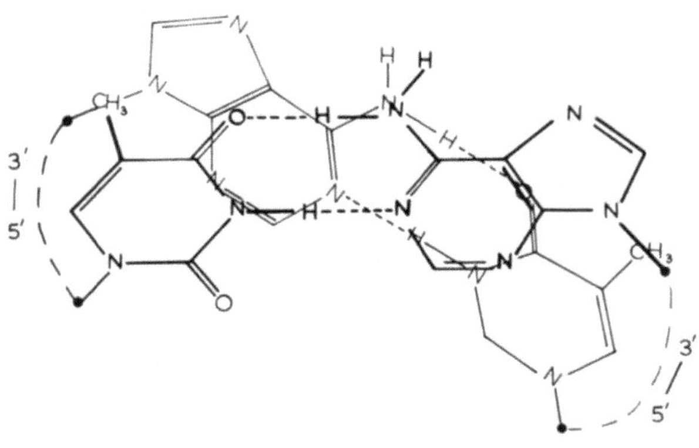

Fig. 17. A diagram of the relative positions of two A—T pairs above one another in DNA. The sequence runs A—T in the 3'—5' direction on both strands. Note the positions of the 6-amino groups on the adenine rings

Robins [26] has recently extended his kinetic studies to investigate the rates of reaction of Pt(^{14}C-en)Cl_2 with polynucleotides containing adenine bases in order to determine whether such an interbase link of the type suggested here can be formed at a significant rate. Second order rate constants for the disappearance of Pt(en)Cl_2 were determined for the compounds shown in Table 2. Although adenine and adenosine react very slowly at 37° C other adenine derivatives react much more rapidly. The order of rate constants is poly A>AMP>ApA>poly-d(AT). Robins [25] observed that the presence of a phosphate group increased the rate 50—60 fold, an observation which parallels his earlier findings. The extremely high rate of reaction with poly-A is attributed to the presence of free terminal phosphate groups. It is especially interesting that the bases in the dimer $A_{3'p5'}A$ and the alternating copolymer poly-d(AT) also react rapidly. The latter will have amino groups of adenine bases on opposite strands well oriented for bi-dentate binding to Pt(en)Cl_2.

In sum the chemistry of the interaction of *cis*- and *trans*-$Pt(NH_3)_2Cl_2$ with nucleotides shows that many sites on pyrimidine and purine bases are open to attack. A difference between the modes of binding of the two isomers can arise only when bi-dentate sites of chelation are available. Adenosine itself furnishes such a site between the N-7 and 6-NH_2 positions. Bi-dentate binding of the *cis* isomer can also take place between groups on two adjacent bases. One example uncovered here is the two amino groups of adenosine rings in the dinucleotide $A_{3'p5'}A$. If such binding occurred in DNA it would introduce a stable link between two bases within one strand or between two bases in opposing strands of DNA. Kinetic experiments confirm that adenine rings can be attacked rapidly when free phosphate groups are available to assist. But, in contrast, these experiments also show a very high rate of reaction of the N-7 position of the guanine ring and this has led one worker to conclude that in DNA itself the guanine bases are first attacked. However, no clear difference between the modes of binding of the *cis* and *trans* isomers to guanosine have yet been demonstrated.

It is not possible from such model studies described here to assess the relative importance of any of these modes of attack for the biochemistry of the platinum drugs. The significant questions as to which bases, *in vivo*, are attacked and at what sites platinum is bound can only be answered by the use of highly sensitive labels to follow and locate the platinum. In the last section of this article we discuss the chemistry of one possible label, a highly coloured platinum complex.

Binding to Amino-Acids and Proteins

The emphasis given in this article to the binding of *cis*- and *trans*-$Pt(NH_3)_2Cl_2$ to nucleic acid components reflects the evidence of the biochemists and microbiologists that attack of platinum on DNA is an extremely important lesion. However it is well known that both isomers will bind to amino-acids and undoubtedly to proteins. At least one anti-tumour compound, Pt(en)Cl_2, has been shown to bind to the sulphur of methionine and used to label proteins for X-ray crystallography. Thus it would not be wise to ignore the possible rate of protein binding in the transport of platinum drugs or indeed in their anti-tumour action or toxicity. A full account of binding to amino-acids and proteins was given by Thomson *et al.* [32]. Essentially no new information has appeared since that review was completed.

"Blue" Platinum Compounds

Derivatives of uracil and thymine will, under certain circumstances, slowly form blue or purple complexes with *cis*-platinum(II) complexes. The details of this interaction are discussed elsewhere in this symposium. Here a short summary is given of the inorganic chemistry, as far as it is known, of blue platinum compounds [3, 6].

Blue complexes of platinum have invited the attention of inorganic chemists because the majority of complexes of platinum, in both oxidation states II and IV, with first row atoms as ligands, are most often yellow or occasionally red. Thus the blue compound, called platinblau, which arises on shaking Ag_2SO_4 with dichloro di-acetonitrile platinum(II), $PtCl_2(CH_3CN)_2$, was regarded as a curiosity [12].

The blue product, which has the composition $Pt(CH_3CONH)_2 \cdot H_2O$, was shown to be monomeric by ebulliometry, an important point since some highly coloured platinum compounds are known which owe their colour entirely to a Pt-Pt interaction arising either in a polymer or in the solid state. The same compound is formed simply by warming the complex $PtCl_2(CH_3CN)_2$. This reaction most likely involves the loss of chloride ion to form the hydrolysis product $Pt(H_2O)_2(CH_3CN)_2$ followed by subsequent platinum catalysed addition of water to the C≡N bond of acetonitrile to form the amide group CH_3—CO—NH_2. Similar complexes have been formed by direct reaction of K_2PtCl_4 with molten amides except those with gem di-N-substituents.

The detailed molecular structure of the blue product, $Pt(CH_3CONH)_2 \cdot H_2O$, is still uncertain, but most work suggests the complex is a bis-bidentate chelate of the amide ion, $CH_3(CO)NH^-$,

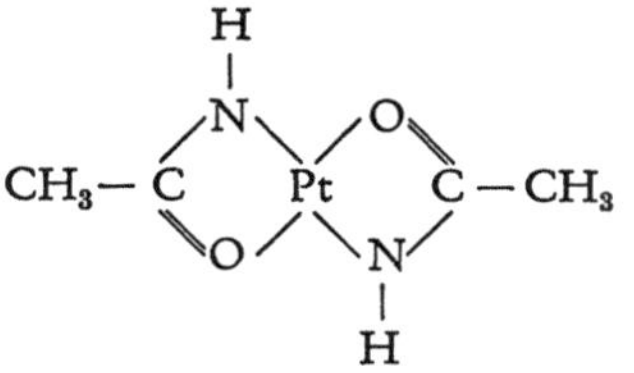

Nuclear magnetic resonance and infra-red evidence supports this conclusion. However, one group of workers formulate the compound as Pt(IV) on the basis of a structure which they suggest is

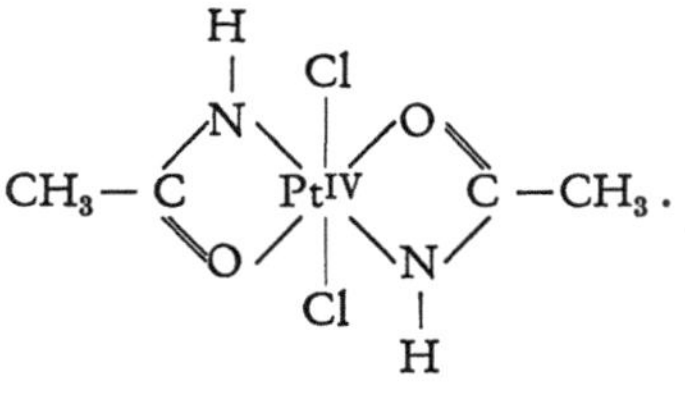

(BROWN *et al.* [3]) although no direct evidence is given for the presence of chloride ion in the product. It now seems unlikely that it is a platinum(IV) complex since it is possible to generate $[Pt(II)Cl_4]^{2-}$ or $[Pt(II)(CN)_4]^{2-}$ simply by the addition of chloride ion or cyanide ion to the platinblau. It is extremely unlikely that these reagents reduce platinblau ([29]). More recently electron spectroscopy (ESCA) measurements have shown that it contains platinum(II).

The ease of formation of blue compounds from $[PtCl_2(CH_3CN)_2]$ suggested its use to prepare compounds of thymidine, uridine and, possibly, of nucleic acids. Preliminary studies by ROOS and THOMSON show that $PtCl_2(CH_3CN)_2$ reacts rapidly with 1-methylthymine to form a blue coloured solution, see Fig. 18. Addition of chloride ion regenerated the original yellow colour of the starting material, $PtCl_2(CH_3CN)_2$, and not the red colour of $[PtCl_4]^{2-}$ showing that the blue colour arose from attack on 1-methylthymine and not from hydrolysis of the co-ordinated acetonitrile. A similar reaction was observed with cytosine but not with guanosine or adenosine.

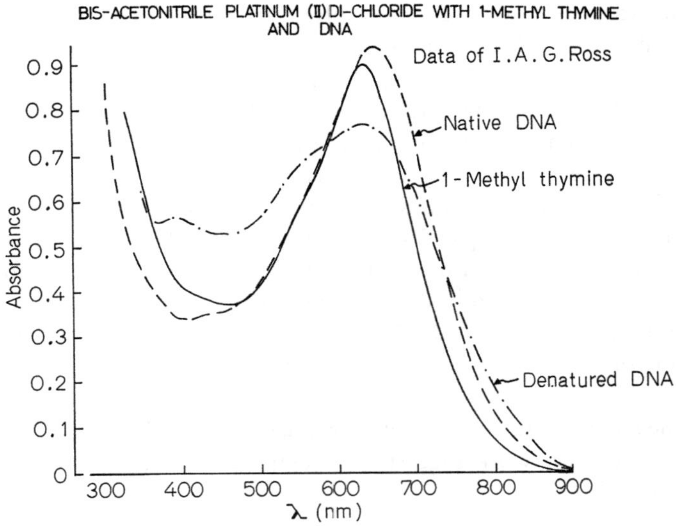

Fig. 18. The absorption spectra (1 cm path) of the blue solutions formed between Pt(II)Cl_2-$(CH_3CN)_2$ and 1-methyl-thymine (———), heat denatured salmon sperm DNA (— · —) and native DNA (— —)

On adding $PtCl_2(CH_3CN)_2$ to DNA a blue colour rapidly formed. The reaction appeared to be complete within an hour or two with denaturated DNA and within 5—6 h with native DNA. Fig. 18 shows the absorption spectra of the resulting species. The speed of this reaction with DNA is very much faster than the rate of reaction of any of the platinum drug analogues we have investigated. The base specificity and the nature of this binding to DNA is clearly of great interest and must now be investigated in detail.

Acknowledgement

is given to the following for allowing the author to see their work prior to publication:

Dr. J. Drobnik, Dr. P. Horacek, Dr. S. Mansy, Dr. A. B. Robins, Professor B. Rosenberg, Mrs. L. Van Camp.

Dr. I. A. G. Roos carried out many of the experiments described here.

References

1. Bannerjea, F., Basolo, F., Pearson, R. G.: Mechanism of substitution of complex ions. XII. Reactions of some platinum(II) complexes with various reactants. J. Amer. chem. Soc. **79**, 4055 (1957).
2. Basolo, F., Pearson, R. G.: Mechanisms of inorganic reactions. New York: Wiley 1958.
3. Brown, D. B., Burbank, R. D., Robin, M. B.: Platinblau. J. Amer. chem. Soc. **91**, 2895 (1969).
4. Day, R. O., Seeman, N. C., Rosenberg, J. M., Rich, A.: A crystalline fragment of the double helix; the structure of the dinucleoside phosphate, Guanylyl-3′, 5′-Cytidine. Proc. Nat. Acad. Sci. (Wash.) **70**, 849 (1973).

5. Gel'fman, M. I., Kustova, N. A.: Complexes of platinum(II) with 5-fluorouracil. Russ. J. Inorg. Chem. (English translation) **15**, 1602 (1970).
6. Gillard, R. D., Wilkinson, G.: Platinum blue and related compounds. J. Chem. Soc. **1964**, 2835.
7. Grinberg, A. A., Panteleeva, E. P.: The acid-base properties of the geometrical isomers of complex compounds. Russ. J. Inorg. Chem. (English translation) **8**, 1165 (1963).
8. Grinberg, A. A., Varshavskii, Yu., S., Gel'fman, M. I., Kiseleva, N. V., Smolenskaya, D. B.: Complexes of platinum(II) and palladium(II) with 6-mercapto-purine. Russ. J. Inorg. Chem., (English translation) **13**, 422 (1968).
9. Green, B., Homer, R. B., McDonough, G. R.: The dAMP-Ag^+ complex and the solubilisation of benzo[a]pyrene. Europ. J. Biochem. **11**, 427 (1969).
10. Gmelins Handbuch. Platin. Teil D. Nummer System 68. Weinheim: Verlag Chemie, GmbH, 1957.
11. Harder, H. C., Rosenberg, B.: Inhibiting effects of anti-tumour platinum compounds on DNA, RNA and protein synthesis in mammalian cells *in vitro*. Int. J. Cancer **6**, 207 (1970).
12. Hofmann, K. A., Bugge, G.: Vergleich der Nitrile und Isonitrile im Verhalten gegen Metallsalze, ein Beitrag zur Konstitution der Doppelcyanide. Berichte **41**, 312 (1908).
13. Horacek, P., Drobnik, J.: Interaction of *cis*-dichlorodiammine platinum(II) with DNA. Biochim. biophys. Acta (Amst.) **254**, 341 (1971).
14. Kondo, N. S., Holmes, H. M., Stempel, L. M., Ts'o, P. O. P.: Influence of the phosphodiester linkage (3'—5', 2'—5' and 5'—5') on the conformation of dinucleoside monophosphate. Biochem. **9**, 3479 (1970).
15. Kotowycz, G., Hayamizu, K.: A carbon-13 NMR study of nucleotide-metal interactions, binding of manganese(II) with adenine nucleotides. Biochem. **12**, 517 (1973).
16. Mansy, S., Rosenberg, R.: Unpublished observations (1971).
17. Mansy, S.: The interaction of *cis*- and *trans*-dichlorodiammine platinum(II) Co-ordination complexes with DNA and its components *in vitro*. Ph. D. Thesis. Michigan State University (1972).
18. Mansy, S., Rosenberg, R., Thomson, A. J.: Binding of *cis*- and *trans*-dichlorodiammine platinum(II) to nucleosides. I. Location of binding sites. J. Amer. chem. Soc. **95**, 1633 (1973).
19. Mogilevkina, M. F., Volshstein, L. M.: Monomethionine platinum(II) chloroplatinates(II). Russ. J. Inorg. Chem. (English translation) **10**, 293 (1965). Further references to the details of this work are given in Thomson *et al.* (1972).
20. Moshkovskii, Yu. Sh., Malysheva, L. F., Mirlina, S. Ya., Seytanidi, K. L.: Selectivity of interaction of potassium chloroplatinite with deoxyribonucleic acid. Biophysics (English translation) **13**, 375 (1968).
21. Pascal, P.: Nouveau traité de chimie minérale. Tome XIX. Masson et Cie., Paris (1958).
22. Reishus, J. W., Martin, D. S., Jr.: *cis*-dichlorodiammine platinum(II). Acid hydrolysis and isotopic exchange of the chloride ligands. J. Amer. chem. Soc. **83**, 2457 (1961).
23. Renshaw, E., Thomson, A. J.: Tracer studies to locate the site of platinum ions within filamentous and inhibited cells of *Escherichia coli*. J. Bact. **94**, 1915 (1967).
24. Reslova, S.: The induction of lysogenic strains of *Escherichia coli* by *cis*-platinum(II) diammine dichloride. Chem.-Biol. Interactions **4**, 66 (1971).
25. Robins, A. B.: The reaction of ^{14}C-labelled platinum ethylenediamine dichloride with nucleic acid constituents. Chem.-Biol. Interactions **6**, 35 (1973).
26. Robins, A. B.: The reaction of ^{14}C-labelled platinum ethylene diamine dichloride with adenine compounds and DNA. Chem.-Biol. Interactions **7**, 11 (1973).
27. Rosenberg, B., van Camp, L., Trosko, J. E., Mansour, V. H.: Platinum compounds: A new class of potent anti-tumour agents. Nature (Lond.) **222**, 385 (1969).
28. Rosenberg, B., van Camp, L.: Reported in these proceedings (1973).
29. Roos, I. A. G., Thomson, A. J., Mansy, S.: Unpublished work (1973). A preliminary report of this work has appeared in the Proceedings of the Seventh International Cancer Chemotherapy Congress. Prague: Avicenum Press. 1971.
30. Roos, I. A. G., Thomson, A. J., Eagles, J.: Mass spectrometric studies of the interaction of platinum complexes with nucleoside analogues. Chem.-Biol. Interactions **8**, 421 (1974).

31. Simpson, R. B.: Association constants of methyl mercuric and mercuric ions with nucleosides. J. Amer. chem. Soc. **86**, 2059 (1964).
32. Thomson, A. J., Williams, R. J. P., Reslova, S.: Complexes related to *cis*-$Pt(NH_3)_2Cl_2$. An anti-tumour drug. Structure and Bonding. **11**, 1 (1972).
33. Truter, M. R., Milburn, C. W.: The crystal structure of *cis* and *trans*-dichloro diammine platinum(II). J. chem. Soc. (A), 1609 (1966).
34. Ulanov, B. P., Malysheva, L. F., Moshkovskii, Yu. Sh.: Determination of the sequence of bases in nucleic acids. Biophysics (English translation) **12**, 371 (1967).
35. Warshaw, M. W., Tinoco, I.: Optical properties of sixten dinucleoside phosphates. J. molec. Biol., **20**, 29 (1966).

Interactions with Biomacromolecules

A. B. ROBINS (Rapporteur)

1. Spectrophotometric study of interaction of *cis*- and *trans*-$Pt(NH_3)_2Cl_2$ with purine and its derivatives.
   V. KLEINWÄCHTER, Institute of Biophysics, Brno, Czechoslovakia.
2. Mass spectroscopic studies of the interactions of platinum complexes with nucleoside analogues.
   I. A. G. ROOS and A. J. THOMSON, University of East Anglia, England.
3. The formation of intra-strand linkage in poly-U via possible Pt-Pt bonds.
   S. MANSY, Purdue University, USA.
4. Interaction of platinum compounds with DNA.
   J. P. MACQUET and T. THEOPHANIDES, University of Montreal, Canada.
5. The interaction of *cis*-$Pt(NH_3)_2Cl_2$ with DNA.
   P. J. STONE and F. M. SINEX, Boston University Medical School, USA.
6. Specific biological activity of platinum complexes. Contribution to the theory of molecular mechanism.
   J. DROBNIK and P. HORACEK, Institute of Macromolecular Chemistry, Prague, Czechoslovakia.
7. Renaturation of DNA by *cis* and *trans* platinum(II) and (IV) compounds.
   H. C. HARDER, George Washington University, USA.
8. *Cis*-dichlorodiammine ^{193m}Platinum binding studies and the effects of binding on the thermal chromicity of synthetic and duplex polynucleic acids.
   H. C. HARDER, George Washington University, USA.
9. The effect of platinum complexes on the structure and function of DNA.
   R. C. SRIVASTAVA, J. FROEHLICH and G. L. EICHHORN, N. I. H. Baltimore, USA.
10. Binding of an antitumour platinum compound to DNA, histones, L-amino acids, poly-L-amino acids, nucleosides and nucleotides.
    C. R. MORRIS and G. R. GALE, Medical University of South Carolina, USA.

Introduction

The contributions to this session have been arranged, in what is considered to be a logical order. That is to say the reactions between the platinum compounds and small molecules have been reported first, and reactions with macromolecules later. All the studies have been carried out on systems at equilibrium, usually after some days have elapsed from the initiation of reaction. From a chemical standpoint, this is meaningful, but it could be maintained that from a biological standpoint, kinetic data would be more relevant, bearing in mind the short period of time over which most drugs are present in an organism at a reasonable concentration. All the papers submitted to this session are concerned with isolated biological macromolecules, or components of such, and such a comment is not intended as a criticism of such investigations.

Table 1. Spectral changes observed on interaction of purine and its derivatives with *cis*- and *trans*-$[Pt(NH_3)_2Cl_2]$

Compound	*cis*-$[Pt(NH_3)_2Cl_2]$			*trans*-$[Pt(NH_3)_2Cl_2]$		
	$\Delta\tilde{\nu}$ [cm^{-1}]	$\Delta\varepsilon$	Shoulder	$\Delta\tilde{\nu}$ [cm^{-1}]	$\Delta\varepsilon$	Shoulder
Purine	−1600	−2300	+	−1100	− 600	+
6-methylpurine	−1100	+ 300	+	−1100	+ 200	*
2-aminopurine	−1600	+2500	○	− 800	+ 900	○
Adenine	−1000	−2000	*	−1300	−2200	(+)
Adenosine	− 200	−1000	*	− 500	−2200	(+)
Deoxyadenosine	− 350	−3500	*	− 500	−2500	(+)
2-methyladenine	− 650	−1500	*	−1000	− 500	(+)
Adenosine monophosphate	− 700	−3400	*	−1050	−2200	(+)
1-methyladenosine	− 600	− 400	*	− 800	−2000	(+)
N6-methyladenine	−1300	−1600	○	−1800	0	○
N6-dimethyladenine	−1200	0	○	−1500	+ 100	○
N6-dimethyladenosine	0	0	○	0	0	○
2,6-diaminopurine	− 600	−1400	○	− 600	−1300	○
2,6-di(diethylamino)purine	0	− 400	○	− 400	− 600	○
Isoguanine	−1100	−1200	○	−1300	−1200	○
Hypoxanthine	−1800	−1600	○	−1900	−1200	○
Inosine	− 400	−2400	○	− 500	−2100	○
Xanthine	− 500	−2100	○	− 300	− 700	○
Theobromine	0	0	○	0	0	○
Caffeine	0	0	○	0	0	○
Guanosine	− 800	−2200	○	− 400	−1600	○
1-methylguanosine	− 700	−2400	○	− 600	−2100	○
7-methylguanosine	0	0	○	0	0	○
Guanosine monophosphate	− 650	−3000	○	− 400	−1800	○

The first author, Kleinwächter allowed 12 days for the establishment of equilibrium between substituted purines and *cis* and *trans* dichloro-diammine platinum(II) (to be referred to as *cis*- and *trans*-Pt(II)). He then investigated the spectral changes, that is any shift in peak wavelength and/or change in extinction upon reaction. All the purines showed a decreased extinction with the exception of 6-methylpurine and 2-aminopurine (Table 1).

The reactions took place in unbuffered, aqueous solution, and it is stressed by the author that the results are qualitative only. In Table 1, a cross in the column headed "shoulder" indicates a shoulder on the low energy side of the maximum, and an asterisk indicates that this shoulder is more pronounced. Caffeine and theobromine, in which the heterocyclic nitrogens are alkylated, and also dimethyladenine and its nucleoside showed no reaction with the platinum compounds. It was found that *cis*- and *trans*-Pt(II) reacted in a similar manner in most cases. Adenine showed an additional shoulder compared with other purines and the suggestion is made that this resulted from a product with stoichiometry Pt : adenine = 2. If the adenine was substituted at N-9, the spectral changes on reaction were lessened — slightly by methyl substitution, more so by ribose. It is interesting that in the guanine series, 7-methylguanosine showed no spectral changes, suggesting that no reaction had occurred. This compound has given conflicting results, as has been pointed out by THOMSON in his review at this meeting.

ROOS and THOMSON have made a study of the mass spectra of methyl-substituted heterocyclic bases (MeB) after reaction with *cis*- and *trans*-Pt(II). In their data, the platinum compounds are referred to as PtA_2Cl_2 where the base A was ammonia,

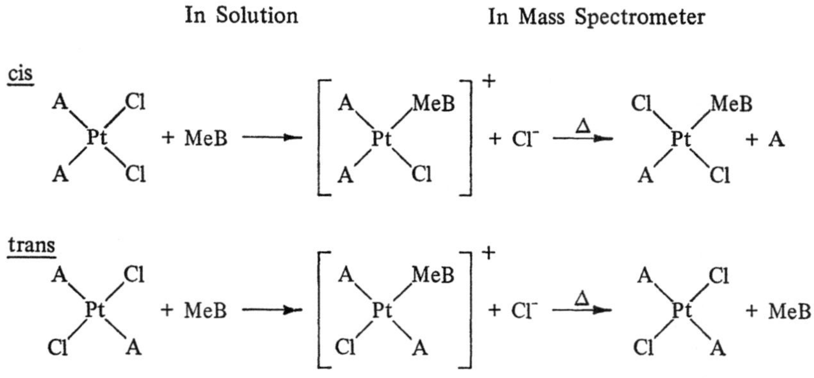

Fig. 1. Reaction scheme for PtA_2Cl_2 and methylated bases in solution and subsequently on the mass spectrometer probe

ethyleneimine, cyclopropylamine or cyclohexylamine. Methylated bases were employed because nucleosides are generally involatile. After the bases had reacted in aqueous solution with equimolar concentrations of the platinum compound for 14 days, the mixture was evaporated on the mass spectrometer probe at 240° C. Under such conditions, the ionic species in solution gave further products which depended on the original reactants (Fig. 1).

For example, *cis*- and *trans*-Pt(II) gave different products with the same base because under probe conditions, the chloride ion labilizes the amine in the *cis* situation, and the methylated base in the *trans*. Thus, from *trans*-Pt(II) reaction the original reactants were reformed. This type of behaviour was observed with 9-methyladenine and 1-methylcytosine which reacted as mono-dentate ligands to give $PtA(MeB)Cl_2$. 1-methylthymine did not react, and 9-methylguanine seemed to react with *cis*-Pt(II) but not with *trans*-Pt(II). The evidence for this rested on the fact that no products were observed under probe conditions. The authors suggest that two chlorine atoms have been replaced by 7-methylguanine acting as bidentate ligand, leading to an involatile product.

The cracking patterns of the compounds of type $PtA(MeB)Cl_2$ can be examined to derive information about the reaction sites of the different bases. The metal-base bond is stable under probe conditions, and thus high m/e values are significant. From such analyses as these, it appears that in the case of 9-methyladenine, N-1 is involved in metallation, and possibly also 6-NH_2. The situation with 1-methylcytosine is more complex but suggests involvement of 4-NH_2. Thus the general conclusions of these authors are that 9-methylguanine reacted in a bidentate manner, 9-methyladenine at N-1 and possibly 6-NH_2 and 1-methylcytosine at 4-NH_2. 1-methylthymine was unreactive.

Although it has been frequently found that uracil and thymine do not react with *cis*- and *trans*-Pt(II), under certain conditions a blue coloration will develop. Such platinum-blue compounds are discussed in other parts of the symposium. Mansy has investigated complexes between *cis*- and *trans*-Pt(II) and polyuridylic acid and polyuridylate-polyadenylate co-polymer. At a concentration of 0.5 Pt to 1 phosphate in poly U, *cis*-Pt(II) produced a blue colour, *trans*-Pt(II) a yellow. The C=O stretch

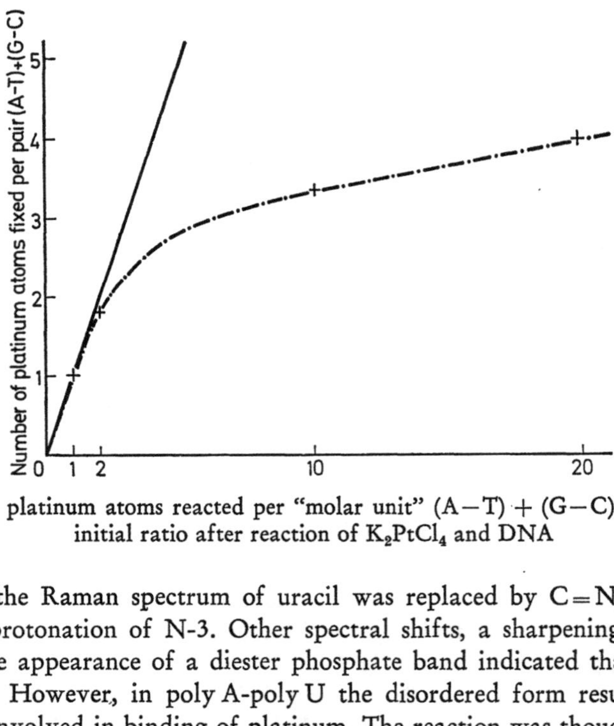

Fig. 2. Ratio of platinum atoms reacted per "molar unit" (A—T) + (G—C) as a function of initial ratio after reaction of K_2PtCl_4 and DNA

frequency in the Raman spectrum of uracil was replaced by C=N, both of these suggesting deprotonation of N-3. Other spectral shifts, a sharpening of all Raman bands, and the appearance of a diester phosphate band indicated that base-stacking had occurred. However, in poly A-poly U the disordered form resulted, and both strands were involved in binding of platinum. The reaction was thought to occur via the aquated form of the platinum complexes since only *cis* diaquodiammine Pt(II) produced a blue colour with uracil derivatives. *Cis*-Pt(II), *trans*-Pt(II) and *trans* diaquodiammine Pt(II) did not behave in this way. Amongst other polynucleotides, poly A, poly C, poly G, poly I and DNA were all disordered by *cis*- and *trans*-Pt(II), and no binding to sugar or phosphate residues was observed. The blue complexes formed from *cis* diaquodiammine Pt(II) and uracil, uridine and thymine showed antitumour activity against ascites 5-180. Complexes with deoxyuridine, 5-fluouracil and 5-bromouracil are to be investigated.

In all subsequent contributions, we shall be dealing with macromolecules and their reactions with the platinum complexes. For the most part, this means intact DNA, although a small amount of data are available on polypeptides and protein.

The first contribution is this group is by MACQUET and THEOPHANIDES, who chose to study the reaction between K_2PtCl_4 and DNA, although as they point out, K_2PtCl_4 is not a biologically relevant compound. It has been used to label DNA for electron microscope studies, and this is mentioned by THOMSON in his contribution. Earlier Russian workers have suggested that guanine and cytosine react preferentially, and a similar conclusion is drawn by MACQUET and THEOPHANIDES. They have analysed the binding of K_2PtCl_4 and DNA in terms of the "molar unit" A-T (59%) + G-C (41%). This represents two base pairs, or four bases, on average. Fig. 2 illustrates the binding as measured by chlorine replacement when increasing amounts of K_2PtCl_4 were added to DNA.

The curve in this figure shows that binding was approximately linear at the ratio of 1 Pt to each molar unit when the reactants were present in the ratio of 2 Pt/molar unit. The final limit in the presence of 100 Pt/molar unit was 4.6 Pt/molar unit bound

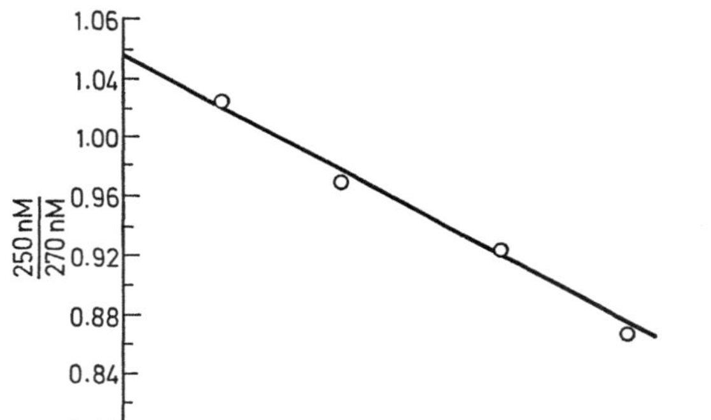

Fig. 3. Ratio of absorbances OD_{250}/OD_{270} as a function of Pt/DNA-P ratio after reaction of DNA and *cis*-Pt(II)

i. e. roughly 1 Pt per DNA base. In the initial stages, up to 0.76 Pt/molar unit produced no change in pH, and the authors suggest that this represents G-C binding since this base-pair is present to 0.82 per molar unit. As binding proceeded, there was an abrupt increase in the pH of 2.7 units which occurred at 2 Pt/molar unit. This corresponded to the liberation of one proton per molar unit, and no further pH change was observed with increase in Pt bound.

The succeeding papers, which involve the reaction of Pt complexes with DNA, concern changes in DNA secondary structure such as melting profiles. These papers will be grouped together because they overlap in some areas.

When DNA is heated in dilute solution the co-operative phenomenon of "melting" occurs, and a sharp increase in optical absorbance (hyperchromicity) occurs over a narrow temperature range. The mid-point of this range is the melting point, T_m. The increase in absorbance at 260 nm may be as much as 35%. There is no shift in the wavelength of peak absorbance during melting. However, the binding of molecules such as the platinum complexes to DNA is accompanied by a shift in the DNA spectrum, and the ratio OD_{250}/OD_{270} has been used as a measure of this [1]. Stone and Sinex employ this parameter in their contribution to this meeting, and Fig. 3 illustrates the sensitivity of this ratio to the degree of binding. The binding was determined by density gradient centrifugation, the platinum atom conferring a higher density to the DNA, and also by a microanalytical technique [1].

It was shown earlier that although DNA did not normally re-nature on cooling, in the presence of 0.05 Pt/DNA-phosphorus renaturation was extensive [1]. This renaturation probably resulted from the presence of a small number of induced interstrand links which maintained the DNA strands in some small degree of register so that the remainder of the base-pairs readily reformed in the original manner on cooling. Stone and Sinex have examined this process more closely, with the results given in Table 2.

Table 2. $\varepsilon/\varepsilon_0$ for various reaction parameters

Reaction ⟶	*Melt* ⟶	*Renature* ⟶	*Melt* ⟶	*Renature*
Pt : DNA = 1 : 4 dialyze after 2 hours at 24° C	1.28	1.03	1.26	1.07
Pt : DNA = 1 : 40 or less dialyze	1.29	1.16	1.26	1.22
Pt : DNA = 1 : 4 don't dialyze	1.26	1.24	1.27	—
Pt : DNA = 1 : 4 melt DNA then add *cis*-Pt	1.40	1.32	1.32	1.32

Complete renaturation is indicated by a value of the hyperchromicity E/E_0 of 1.0, and de-naturation by a value of approximately 1.3. In the presence of 1 Pt to 4 DNA-P a considerable degree of renaturation occurred after 2 hours incubation of the reactants, and dialysis of unreacted *cis*-Pt(II) (Line 1, Table 3). When the Pt/DNA-P ratio was lowered by an order of magnitude, much less renaturation resulted (Line 2). Finally we see in Lines 3 and 4 that a large Pt/DNA-P ratio led to lack of renaturation if dialysis was not carried out or if the *cis*-Pt(II) was added after melting. This is presumably the result of reaction between *cis*-Pt(II) with DNA in the single-stranded form such that renaturation of the modified base-pairs was prevented. As reaction of DNA with *cis*-Pt(II) proceeded, the ratio OD_{250}/OD_{270} fell (Fig. 3). This fall occurred more rapidly than corresponding changes in hyperchromicity. Circular dichroism measurements also demonstrated that pronounced changes in configuration occurred early, and denaturation only later. The process was associated with an activation energy of 7—9 K Cal/mole.

There are three other interesting points in this paper. Firstly, the binding of platinum was suppressed by increased chloride ion concentration (Fig. 4) which suggested that prior aquation of the platinum complex was required, and that the equilibrium was shifted by Cl^-. Binding of *cis*-Pt(II) still occurred however, even in the presence of 2.4 M NaCl, and once reaction between *cis*-Pt(II) and DNA had occurred subsequent changes in Cl^- concentration had no effect.

Secondly, chromatin appeared to bind *cis*-Pt(II) less readily than its constituent DNA, which may indicate a shielding by protein. Thirdly, a DNA of high G-C content reacted with *cis*-Pt(II) to produce a *hypo*chromic shift, suggesting an increase

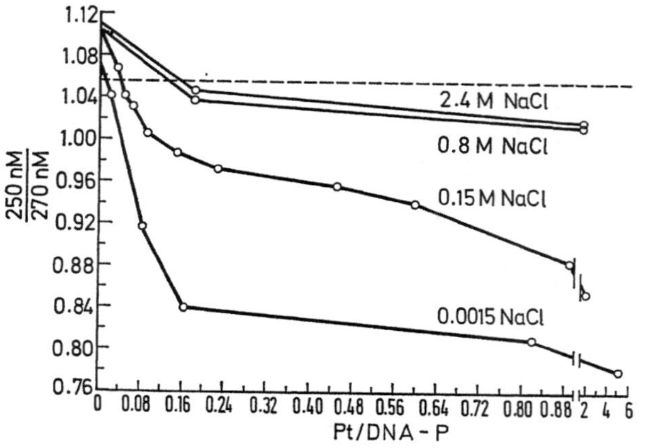

Fig. 4. Equilibrium values of OD_{250}/OD_{270} as a function of the ratio Pt/DNA-P and sodium chloride concentration

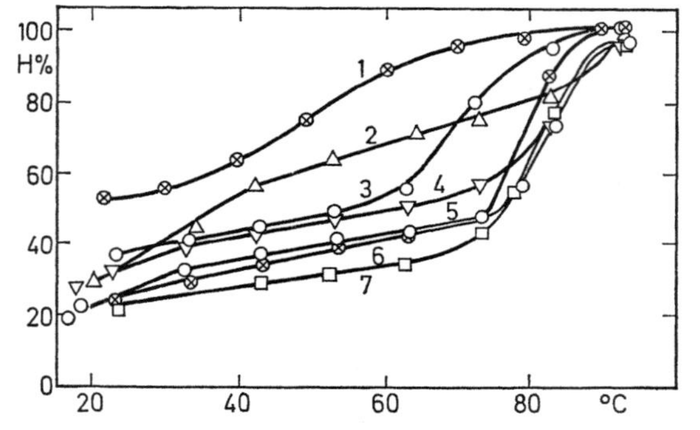

Fig. 5. Renaturation curves of calf thymus DNA in solutions of 0.1 M NaCl and hydrolysed *cis*-Pt(II). Denatured at 93° C for 20 mins, renatured by cooling at 0.7° C/min. Figures apply to varying Pt/DNA-P ratios. Curve (1), Pt/DNA-P = 2.0; (2), 0.0; (3), 1.0; (4), 0.06; (5), 0.13; (6), 0.52; (7), 0.26

in helical configuration. This high G-C DNA showed optimal binding at a ratio of Pt/DNA-P = 0.13, and a low G-C DNA at a ratio Pt/DNA-P = 0.05 which indicated a preferential binding to G-C base pairs.

The papers by Horacek and Drobnik and by Harder are similar and somewhat overlapping studies. Horacek and Drobnik have extended their earlier studies on *cis*-Pt(II)-induced renaturation of DNA [1]. Other *cis* compounds have been investigated and also *trans*-Pt(II). Their main conclusion here is the necessity for the platinum complex to be hydrolysed before reaction with DNA occurs. Such complexes include *cis*-Pt(II), dichloroethylene diamine Pt(II), oxalatodiammine Pt(II) and malonatodiammine Pt(II). *Trans*-Pt(II), K_2PtCl_4 and K_2PtCl_6 have little or no effect or react much more slowly. In the absence of Cl^-, the optimal Pt/DNA-P ratio for renaturation was 0.05 [1]. Increasing the Cl^- concentration suppressed this renaturation, and the optimal Pt/DNA-P rose to 0.25 (Fig. 5, Curve 7).

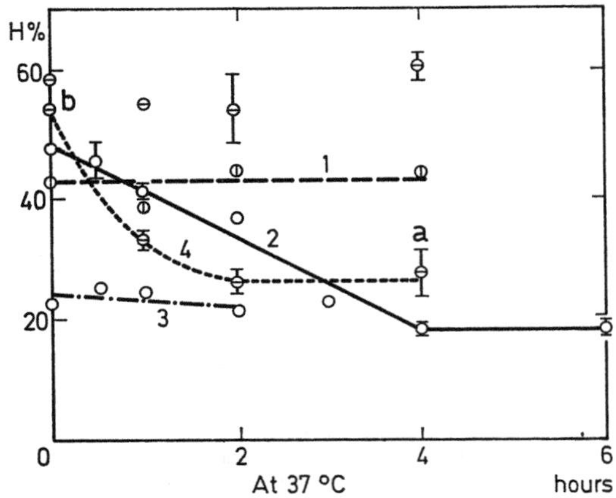

Fig. 6. Hyperchromicity after rapid cooling of denatured DNA solutions. Denaturation followed reaction at 37° C between DNA and *cis*-Pt(II) or *trans*-Pt(II) for varying periods of time. Curve 1. "aged" *cis*-Pt(II)/DNA-P = 0.01; 2. "fresh" *cis*-Pt(II)/DNA-P = 0.05; 3. "aged" *cis*-Pt(II)/ DNA-P = 0.05; 4. "aged" *trans*-Pt(II)/DNA-P = 0.05

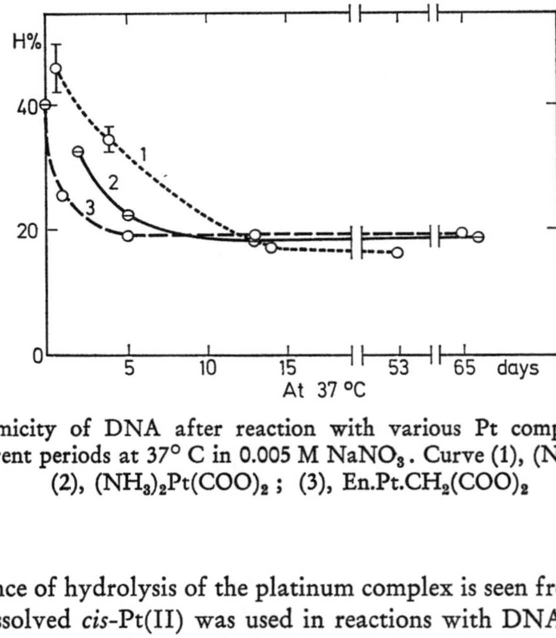

Fig. 7. Hyperchromicity of DNA after reaction with various Pt complexes pre-incubated ("aged") for different periods at 37° C in 0.005 M $NaNO_3$. Curve (1), $(NH_3)_2PtCH_2(COO)_2$; (2), $(NH_3)_2Pt(COO)_2$; (3), $En.Pt.CH_2(COO)_2$

The importance of hydrolysis of the platinum complex is seen from Fig. 6.

If freshly dissolved *cis*-Pt(II) was used in reactions with DNA during a denaturation-renaturation cycle, then marked changes in hyperchromicity resulted over a period of time (Fig. 6, Curve 2) whereas this was not observed with previously equilibrated (hydrolysed?) solutions (Fig. 6, Curves 1 and 3). Once reaction had occurred between the platinum compounds and DNA at any Cl^- concentration, dialysis to remove Cl^- had no effect. This was observed by Stone and Sinex also. Dichloroethylenediamine Pt(II) behaved in a very similar manner to *cis*-Pt(II), but the malonato, and oxalato compounds took many days to attain equilibrium if this was measured by the degree of hyperchromicity ultimately reached (Fig. 7).

These authors suggest that although hydrolysis of the platinum compound seems to be a prerequisite of reaction, the biological environment contains Cl^- which would suppress reaction. Possibly within the cell the platinum complexes, which have entered in an uncharged form, suffer hydrolysis and subsequently react. The platinum complexes with organic ligands (e. g. malonato) hydrolyse more slowly, and may be less toxic for this reason.

One puzzling feature of the complexes is the difference between *cis* and *trans* isomers. The *cis* compounds seem to be biologically active, the *trans* are very much less so. Reactions *in vitro* show both isomers to be closely similar however. HARDER has investigated the renaturation of DNA after reaction with *cis*-Pt(II), *cis*-Pt(IV), ethylenediamine Pt(II), *trans*-Pt(II) and *trans*-Pt(IV). The conditions he used were very similar to those of HORACEK and DROBNIK. Thus, DNA was incubated 14—16 h with

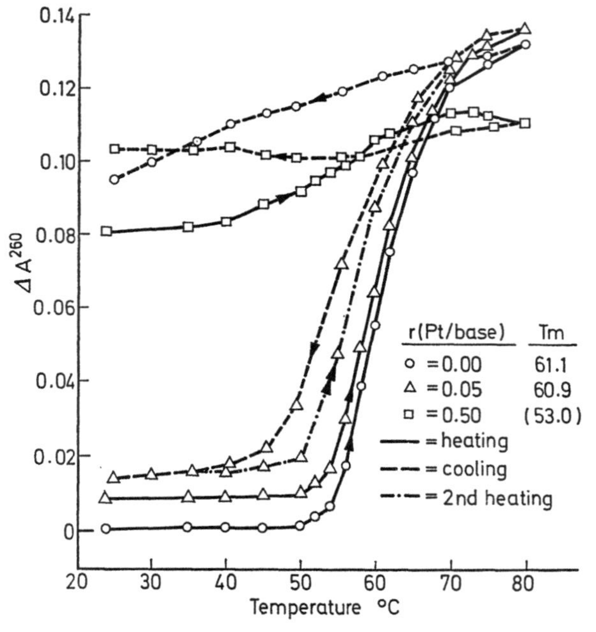

Fig. 8. Denaturation-renaturation curves for calf thymus DNA treated with *cis*-Pt(II). -○-, Pt/DNA-P = 0.00; -△-, 0.05; -□-, 0.50

the platinum compound and then a melting profile determined with slow heating and cooling. T_m values were found also by heating to 95° C followed by quick cooling. Such thermal denaturation-renaturation curves are illustrated in Fig. 8 in which the optimal renaturation is seen to have occurred at Pt/DNA-P = 0.05.

There was some slight renaturation at a ratio Pt/DNA-P = 0.5 but this was "non-cooperative". On further investigation into the renaturation as a function of time and Pt/DNA-P ratio, this is seen even more clearly (Fig. 9).

If the Pt/DNA-P ratio was too high, renaturation was lost, and permanent denaturation resulted (see also STONE and SINEX, Table 2, Line 3). The renaturation produced by Pt-DNA linkages was found to be resistant to acid and alkali treatment.

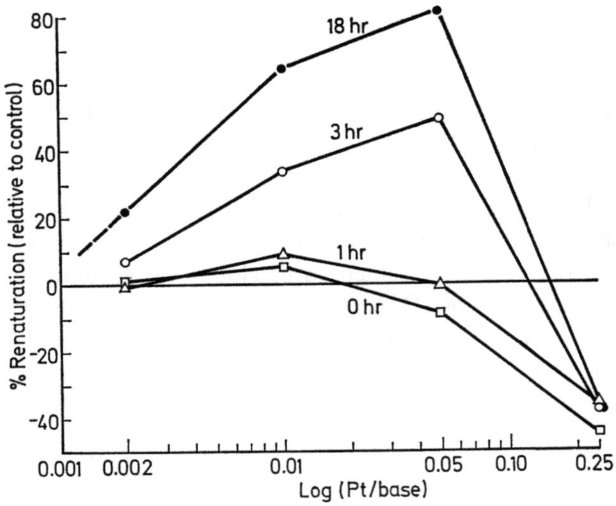

Fig. 9. Degree of renaturation produced by reaction of *cis*-Pt(II) with DNA at different times and ratios Pt/DNA-P

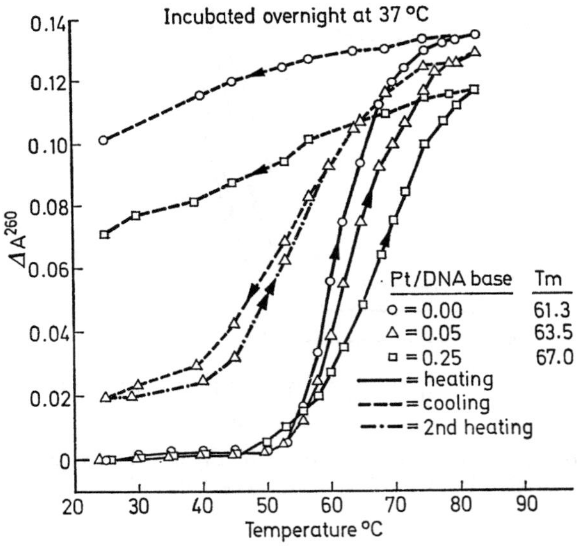

Fig. 10. Denaturation-renaturation curves of DNA treated with *trans*-Pt(IV). –○–, Pt/DNA-P = 0.00; –△–, 0.05; –□–, 0.25

It has been already stressed that *cis*- and *trans*-Pt(II) produce similar chemical effects, and Harder has also found that *trans*-Pt(II) and *trans*-Pt(IV) produce similar renaturation to the *cis* compounds, though somewhat less (Fig. 10).

In most cases, the renaturation phenomenon could be observed by adding DNA to a previously aged solution of the platinum compound, which is further evidence for prior aquation before reaction. This was not true, however, in the instance of *cis*-Pt(IV), and even when the platinum compound and DNA were reacted for 16 h at 20° C, no renaturation resulted — a unique result (Fig. 11).

Only if the *cis*-Pt(IV) was kept in solution for 24 days at 4° C did it produce renaturation of DNA, and then not immediately, and at an optimal Pt/DNA-P ratio of 0.25 (Fig. 12).

HARDER suggests that possibly the DNA itself is necessary to activate, or stabilize the aquated species in the *cis*-Pt(IV). There were some differences between the spectra of *cis*-Pt(IV)-reacted DNA and the products of reaction between DNA and other platinum compounds. To explain data such as those in Fig. 9 where a denaturation

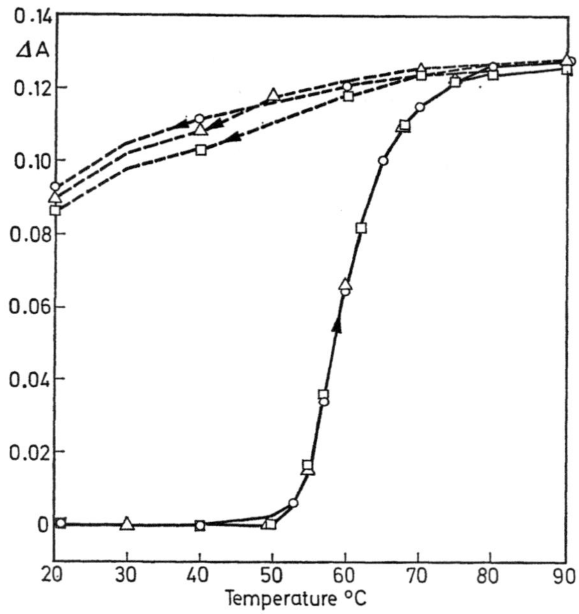

Fig. 11. Denaturation-renaturation curves for DNA treated with *cis*-Pt(IV) for 16 h. –○–, Pt/DNA-P = 0.0; –△–, 0.05; –□–, 0.25

first occurred, followed by renaturation, HARDER suggests that one may have to propose two types of reaction. One of these is rapid and dominates at high platinum concentration, destabilizing the DNA helix. The second process occurs slowly at a low platinum complex concentration and produces renaturation (however this is hard to reconcile with Table 1, Line 1, where only 2 h reaction was allowed and maximum renaturation resulted).

In a second communication, HARDER has described the thermal denaturation in synthetic polynucleotides after reaction with *cis*-Pt(II). No renaturation was observed with poly I-poly C, poly A-poly U, or poly dA-poly dT, and *cis*-Pt(II) although the T_m values were lowered. With poly d(A-T) which renatures completely in the pure state, *cis*-Pt(II) produced a decrease in hyperchromicity, and appreciable loss of the intrinsic renaturation (Fig. 13).

Trans-Pt(II) produced very similar results to *cis*-Pt(II) with this polymer, but *trans*-Pt(IV) was different in that it caused less inhibition of hyperchromicity (it may indeed not react).

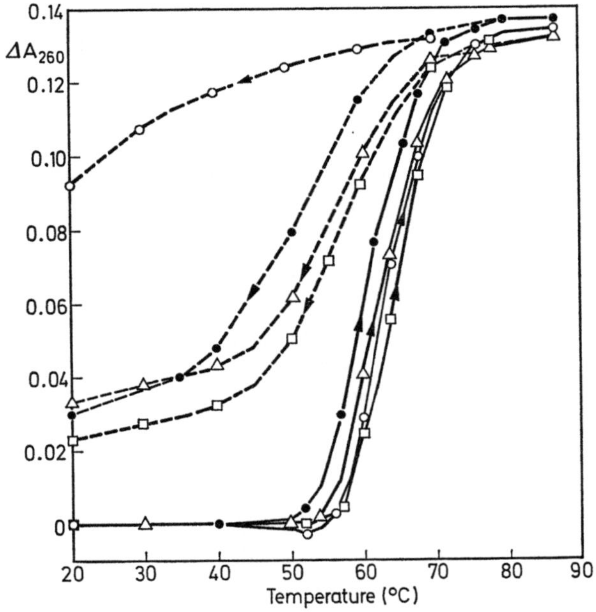

Fig. 12. Denaturation-renaturation curves for DNA treated with *cis*-Pt(IV) for 24 days at 4° C. -○-, Pt/DNA-P = 0.00; -△-, 0.05; -□-, 0.25; -●-, 1.00

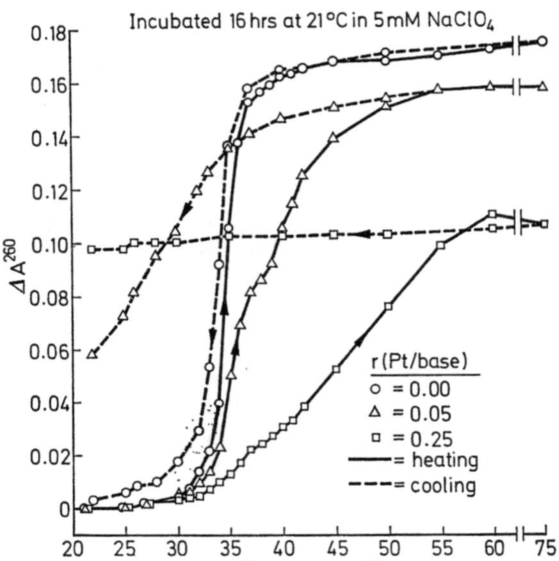

Fig. 13. Denaturation-renaturation curves for polyd(AT) treated with *cis*-Pt(II). -○-, Pt/DNA-P = 0.00; -△-, 0.05; -□-, 0.25

Binding of the *cis*-Pt(II) complex to these polynucleotides was measured by HARDER using ^{193m}Pt as a tracer, and estimating this in acid-precipitated polynucleotide after reaction for various times and at various Pt/base ratios (Table 3).

Table 3. Moles *cis*-$^{193m}Pt(II)(NH_3)_2Cl_2$ bound per moles base nucleotides at 25° C with different incubation molar ratios

After 16 hours	r_b		
	$r_i = .05$	$r_i = 0.25$	$r_i = 0.75$
DNA	0.020	0.077	0.136
poly dAT : poly dAT	0.010	0.050	0.137
poly dG : poly dC	0.030	0.125	0.246
After 19 days			
DNA	0.046 [a]	0.200	0.356
poly dAT : poly dAT	0.041	0.204	0.410
poly dG : poly dC	0.050	0.246	0.444

[a] After 3 days only.

The extent of binding was greatest in the G-C copolymer and least in the A-T over the first 16 h, though such differences disappeared over a longer time interval. There is thus further evidence of preferential reaction at G-C sites in the initial stages.

The contribution by SRIVASTAVA, FROEHLICH and EICHHORN attempts to throw some light on the much-debated subject of *cis* and *trans* compounds, and their different biological effectiveness. These authors re-emphasise that platinum may bridge two adjacent bases in the same DNA molecule in a manner which is depicted in Fig. 14, either in a monofunctional or a bifunctional manner.

However, they point out that a further possibility exists, since *cis*-Pt(II) is present in crystalline form with a Pt-Pt distance of 3.4 Å (Fig. 15). This distance is greater in the *trans*-Pt(II).

Thus *cis*-Pt(II) could co-stack with DNA very exactly, and by examining closely the stoichiometry of reaction products, one could differentiate between these mechanisms (i. e. mono- or di-functional) since co-stacking necessitates 1 : 1 stoichiometry (Fig. 14). Circular dichroism measurements of *cis*-Pt(II)-DNA reaction products of varying reactant ratios showed no evidence of Pt/base 0.5 : 1 stoichiometry but only Pt/base 1 : 1 (Fig. 16).

Trans-Pt(II) gave evidence of 0.5 : 1 stoichiometry with some 1 : 1. Possibly therefore co-stacking could explain the biological activity. However, the authors further found that at low ratios of Pt/base, circular dichroism increased to values above that of DNA itself, followed by a decrease subsequently at higher Pt/base ratios (Fig. 16). This behaviour was not seen in the case of the *trans*-Pt(II) reactions, nor with *cis*-Pt(II) and denatured DNA. The authors suggest that in view of these results and the fact that there was a maximum in the curve at a Pt/base ratio of approx. 1 : 9, this may be the biologically relevant mechanism. On testing the level of RNA synthesis

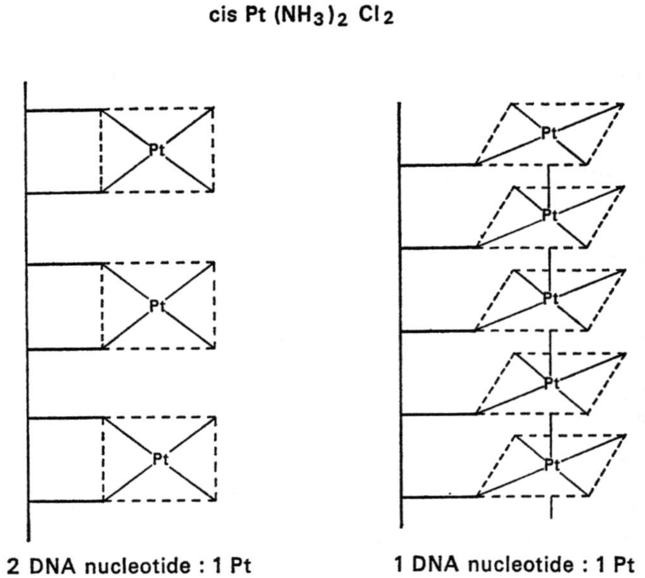

Fig. 14. Possible stereoselective interactions of *cis*-Pt(II) and DNA

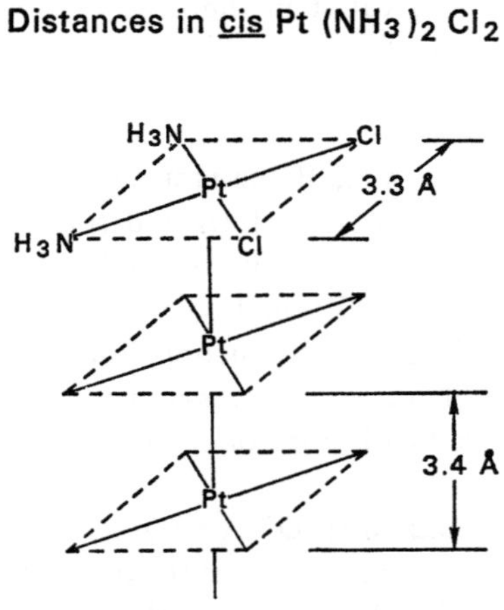

Fig. 15. Interatomic distances in *cis*-Pt(II)

in an *in vitro* system using *E. coli* RNA polymerase and a DNA template previously reacted with *cis*-Pt(II) at a Pt/base ratio of 1 : 9, they found a decrease in synthesis when the template was exposed to *cis*-Pt(II).

It is perhaps strange that very little investigation has been made into the reaction between the platinum complexes and proteins and their constituents. The last paper submitted to this meeting is such a study, including nucleotides, nucleic acids, amino acids and histones. The authors, Morris and Gale have used *cis* dichloro di-(H^3-pyridine) Pt(II) and studied the binding of this compound to DNA by equilibrium dialysis. The retention of the platinum compound by DNA in the dialysis bag was

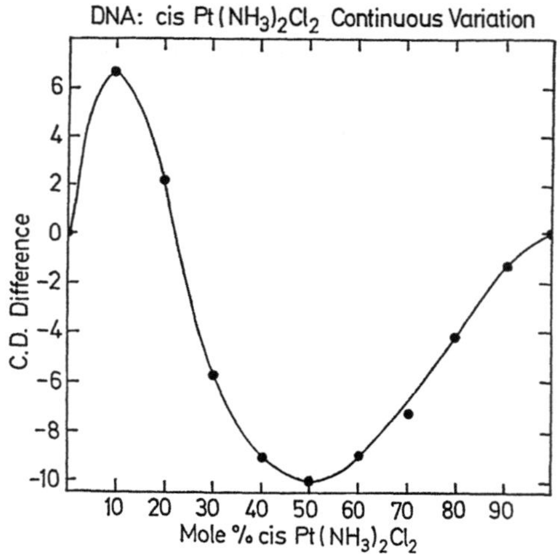

Fig. 16. Circular dichroism difference in DNA-*cis*-Pt(II) reaction mixtures as a function of fractional concentration of *cis*-Pt(II). Total concentration of both components was constant at $6 \cdot 10^{-5}$ M

decreased by prior exposure of the metal complex to amino acids and nucleotides. The results are internally comparable but it must be remembered that binding was not proved, and that what was measured was an association within the dialysis compartment. The greatest inhibition of binding was produced by pre-incubation of the platinum compound with glutamic and aspartic acids, lysine, arginine and cysteine. Histone also decreased subsequent binding to DNA by 50%; the positively charged polymers polylysine and polyarginine repelled the metal compound from the dialysis sack. On the other hand, binding of *cis*-dipyridine Pt(II) occurred readily to polyglutamate and polyaspartate. The authors suggest that the platinum compounds bind preferentially to electron rich centres such as carbonyl and heterocyclic nitrogen in DNA or possibly the acidic proteins which have been implicated in gene control.

The binding of platinum compounds to proteins is considered by this rapporteur to be of considerable importance in studies involving cells in culture or whole organisms. Thus the effectiveness of a compound such as dichloroethylenediamine Pt(II)[Pt(enCl_2)] inhibition of cell growth is lost within a short time because of binding of the drug to the serum proteins of the culture medium. This binding is quite rapid, with a half life of about 50 minutes, but does not appear to be very strong, as it is acid labile. It differs in kind from the binding to nucleic acid and its constituents. Early experiments have shown that the binding is not specific for any one class of protein. The only effect on the protein at low Pt/protein ratios is to alter its electrophoretic mobility. The amount of platinum compound used here was very small — in the biologically realistic region of 1 platinum atom per 5 or 10 intact protein molecules.

The rapporteur would like to conclude by underlining the need for more investigation into the entry of the platinum compounds into cells, and their pharmacology. It seems that when reactions with DNA are examined, in the early stages probably G-C sites are preferentially attacked, which is an indication of the importance of kinetic studies, because such selectivity can disappear over longer time intervals or at high Pt/base ratio. The question of nucleotide pools and their reactivity towards the platinum compounds has not been studied, and very little has been carried out on the proteins and their constituents. However, the chemistry of the interaction of these complexes with the nucleic acid components or intact DNA has been well covered, and the general picture has become clearer.

References

1. Horacek, P., Drobnik, J.: Biochim. biophys. Acta (Amst.) **254**, 341 (1971).

Bacterial, Viral and Tissue Culture Studies on Neutral Platinum Complexes

J. J. ROBERTS

A reviewer of a subject tends to be convinced that everyone is familiar with the interesting and relevant aspects of his topic. The attitude I have taken is that any really good story can withstand any amount of retelling. The discovery of the anti-tumour platinum compounds is just such a story and remains many years after it was first announced as one of the classical examples of serendipity operating in science. Who could have foretold that the modest opening sentence of the first paper which related to these compounds ([14]) and reads: "In an investigation of the possible effects of an electric field on growth processes we have discovered a new and interesting effect" would have been followed by so many other valuable discoveries and result in the innovation into cancer chemotherapy of an entirely new class of chemical agent. In this survey of the bacterial, viral and tissue culture effects of platinum compounds I will begin by briefly describing some of the elegant series of experiments by ROSENBERG *et al.* [14], then mention other relevant studies in the bacterial field, before proceeding to discuss more recent viral and tissue culture studies.

As it was soon appreciated that the platinum compounds had many properties in common with the classical difunctional alkylating agents, I will attempt to point out the similarities and dissimilarities which exist between these two types of agent when this is apparent. Inevitably DNA has figured prominently as a possible target for both these agents in preference to reaction with other macromolecules such as RNA or protein; so I shall try to make it clear when the evidence supports this contention.

The first observation made by ROSENBERG *et al.* [14] was that when a low alternating current is passed through platinum electrodes in nutrient media it can inhibit cell division in *E. coli* and cause the formation of long filaments. It was subsequently discovered that some of the platinum dissolves under these conditions to give first of all the ionic species, ammonium hexachloroplatinate. This compound could, at high concentrations, inhibit cell division. However it was soon noticed that aged solutions of this compound were far more efficient in producing filaments, but only when the solutions were exposed to light. The fresh solutions were a light yellow colour while aged solutions were almost colourless. All this implied that a photochemical change of the platinum complex had occurred and that the products of this agent were more active in inhibiting cell division than the original compound. The changes which occurred in a solution of the hexachloroplatinate were soon elucidated and shown to involve the replacement of the chloride ligands by NH_3 with the loss of one negative charge per replacement to give finally a stable neutral species. This new species was then shown convincingly to inhibit cell division but not growth, which was in contrast to the parent ionic species, which was a bacteriocide and not a bacteriostat. The differences between the various charged and uncharged platinum species are summarised in Table 1.

Salts of other group VIII B transitions metals such as rhodium and ruthenium have been shown similarly to be able to produce filamental growth in *E. coli* but in all cases a much higher concentration was required than with the platinum salt [15].

If grown on nutrient media, almost all the gram negative bacilli examined showed filamentous growth. Some of the gram positive bacilli also showed slight elongation but only at near toxic levels of platinum. On the other hand all the cocci tested showed no noticable effect even at relatively high concentrations of platinum [15]. The difference in the response of the gram positive bacteria was not due to any failure of the platinum compound to penetrate the cell walls as was shown by RENSHAW and THOMSON [16] from studies using radioactive platinum compounds. These authors did, however, observe a difference in distribution of the platinum compound as between gram positive and gram negative organisms. There was a preponderance of the radioactive material in the metabolic intermediates of the gram positive species which suggests that the platinum is retained in the pool, possibly by metabolites complexing with platinum.

Table 1. Summary of effects of platinum complexes on *E. coli* growth

Type of complex	Effects on cell growth	Effects on cell division
Doubly negative, $[PtCl_6]^-$	Inhibitory (bacteriocide) in low concentrations (>1 ppm)	No detectable effect
Singly negative, $[PtCl_5(NH_3)]^-$	No inhibition	Slight inhibition [a]
Neutral, $[PtCl_4(NH_3)_2]^0$	No effect at low concentrations (<4 ppm); slight inhibition at higher concentrations	Inhibitory (filamentous growth) in low concentrations (>2 ppm)
Singly positive, $[PtCl_3(NH_3)_3]^+$	Not tested in sufficient concentration	Not tested in sufficient concentration

[a] The slight inhibition observed may be due to lability of the complex.

The distribution of platinum ions within *Escherichia coli* after the induction of filaments with *cis*-$Pt(II)(NH_3)_2Cl_4$ and after growth inhibition with $PtCl_6^{2-}$ was also determined by these workers using ^{191}Pt which has a half-life of 3 days. In the filamentous cells platinum ions were associated not only with metabolic intermediates but also with cytoplasmic proteins and nucleic acids whereas in the inhibited cell the platinum was combined only with the cytoplasmic proteins. This observation therefore provided the first indication that the filament forming ability of these compounds was associated with nucleic acid interactions.

Induced filamentous growth occurs in response to a variety of agents, for example, U.V.- and X-irradiation or alkylating agents. In many bacteria or *Erwinia* this can be reversed and cytokinesis (cross-septation) initiated by exposure to such agents as divalent metals, L-pantoyllactone or incubation at 42° C. Cytokinesis, or cell division in platinum-induced filaments could only be initiated by removing or decreasing the concentration of the platinum complex. Under these conditions cross-walls formed at intervals along the entire filament, but in general it was found that active cell division occurred near one or both ends of the filament. These observations indicated that the specific platinum-induced inhibition of cell division leading to filament formation in *E. coli* differs from that which occurs in response to other treatments. A key question, however, in this context is whether all the so-called daughter cells are viable with a full complement of normal DNA.

The mechanism of filament formation in *E. coli* is by no means clearly understood, but it is probably true to say that it can be regarded as a manifestation of the ability of a compound, or an agent, to react with DNA such as to lead to an inhibition of DNA synthesis with no accompanying inhibition of RNA or protein synthesis. The accumulation of areas of strikingly enhanced electron density within platinum-induced filaments of *E. coli* were thought by Howle and Gale [4] to be aggregates of ribonucleoprotein which had lost its usual distribution pattern but retained some degree of biochemical integrity, and unimpeded rate of synthesis.

Evidence that DNA can be a target for these neutral platinum salts in *E. coli* is indicated by some recent elegant studies by Drobnik *et al.* [1]. Currently much interest is focussed on the ability of cells to repair damage to cellular DNA. Specific mechanisms for repairing DNA are now known to be determined by various genes in *E. coli*. The response of various strains of *E. coli* which have been mutated at these specific loci can indicate, not only that an agent has reacted with DNA, but also which repair mechanism can recognise the resulting DNA lesion. Drobnik *et al.* [1] therefore studied the effects of *cis*-dichlorodiammineplatinum(II) (*cis*-Pt(II)) on the colony-forming ability of strains of *E. coli* mutated at these different loci. These were the *fil*, *exr.* and *hcr* loci and these were studied in both growing and resting cultures. These loci were known to determine the sensitivity of cells to U.V.-irradiation and to alkylating agents. Mutation in the *fil* marker had a pronounced effect on a *cis*-Pt(II) treated growing culture where it affected both filament formation and the preservation of the colony-forming ability. Thus the mutant *E. coli B* containing the fil^{+} marker formed longer filaments and was more sensitive than the *E. coli* B/r strain which was fil^{-}. The difference was not observed in the resting culture.

The *hcr* locus is known to control an excision repair mechanism in *E. coli*. It determines the excision of U.V.-induced photoproducts, the so-called thymine dimers formed by linking together two adjacent thymine molecules on the same strand of DNA, from *E. coli* DNA and is important for survival after U.V.-irradiation. However this locus does not play a role in the survival after treatment with the alkylating agent *N*-methyl-*N'*-nitro-*N*-nitrosoguanidine (MNNG). The *hcr* locus was found by Drobnik *et al.* [1] to be important to some extent in determining sensitivity to *cis*-Pt(II) but less important than for sensitivity to U.V. suggesting that excision of platinum molecules may not be an important repair process in *E. coli*.

The biological role of the *exr* locus is not completely understood, but it seems to be important in controlling the function that allows replication of DNA, which still contains unexcised thymine dimers. In this model of DNA repair, the newly-synthesised daughter strand DNA probably contains gaps opposite the photoproducts. These gaps are then subsequently filled-in, possibly by *de novo* DNA synthesis. Mutation in the *exr* gene was found to produce a very great increase in the sensitivity of *E. coli* to *cis*-Pt(II). Its role was much more important in determining sensitivity to the platinum compound than to either U.V.- or X-irradiation or the alkylating agent, MNNG. These results therefore strongly support the view that the lethal action of the platinum compound is mediated by reaction with DNA and in this respect it resembles the other agents I have mentioned: U.V.- and X-irradiation and alkylating agents. Moreover the damage introduced into DNA by *cis*-Pt(II), like that introduced by these other agents is recognised by specific DNA repair processes. However the differences in the levels of sensitization to these agents by mutation at the specified markers does indicate that the platinum compound has a specific biological effect in this system, which differs from that produced by either irradiation or alkylating agents.

There is good evidence that these repair mechanisms in *E. coli* have their counterparts in higher organisms and in mammalian cells. This being so, it seems likely that the sensitivity of tumour cells to platinum compounds and other toxic agents which react with DNA, such as the alkylating agents will not always run in parallel, so that some selective action is to be anticipated for these agents.

DROBNIK *et al.* [1] also noticed that the inactivating effect of *cis*-Pt(II) occurred only after several apparently normal cell divisions and they felt this indicated that environmental factors which affect growth and biochemical reactions also determined the efficiency of these DNA repair processes.

I will now consider studies involving simpler systems such as bacteriophage or transforming DNA. As referred to earlier, filament formation in *E. coli* can be regarded almost as a diagnostic indication that a compound has an action on the nucleus of a cell. The term "radiomimetic" was coined to describe compounds which, like X-irradiation, have this property and which, as a consequence of this, elicited a whole range of common biological and biochemical effects. The ability to induce prophage in lysogenic strains of bacteria can, like filament formation, be regarded as indicative of this radiomimetic potential of a compound. Moreover, HEINEMANN and HOWARD [3], from an extensive study of prophage induction with a wide range of chemicals, showed that there is a good correlation between anti-tumour activity and the induction of prophage. RESLOVA [10] has investigated the ability of platinum compounds to induce lysogenic strains of *E. coli*. Two strains of cells were studied, one carrying the defective prophage T15, the presence of which could be verified by electron microscopy, and the other carrying the prophage lambda. A general relationship was found to exist between the ability of a platinum compound to inhibit cell division and induce filamentous growth and its ability to cause induction of phage. Some exceptions existed, however, and an even better correlation was found between the anti-tumour activities of these chemicals and their abilities to induce prophage.

There have been other studies on the interaction of platinum compounds with viruses which have also further indicated the relative importance of reactions with DNA as against those with protein in producing biological effects. KUTINOVA *et al.* [7] demonstrated the inactivation of the infectious activity of extracellular papovavirus SV40 by *cis*-dichlorodiammine platinum(II). Some clues as to the mechanism of inactivation of viruses can be gained by following the capacity of the inactivated viruses to induce either tumour or viral antigens. These authors found that the ability of the virus to induce tumour antigens was inactivated at a slower rate than either the infectivity or the ability to induce viral antigens. These findings corroborate studies in which SV40 or polyoma virus were inactivated by radiation and further indicate that the viral DNA and not the protein coat is the primary target for both irradiation and the platinum compounds. Another study on the inactivation of *B. subtilis* transforming DNA by platinum compounds was likewise a further attempt to evaluate the effect of these agents on the biological function of DNA [9]. By examining the changes in transformation frequency of three unlinked genetic markers, the positions of which on the *B. subtilis* chromosome were known, RESLOVA *et al.* [9] hoped for some indication of specificity in the reaction of the platinum complex with regions of DNA or individual DNA bases. The adenine marker, for example, is localised at the region close to the origin of replication and is rich in guanine-cytosine base pairs in DNA and this was found to be appreciably more sensitive than the methionine marker which is localised at the end of the *B. subtilis* chromosome. A comparative study to this using mono- and difunctional alkylating agents would clearly be of considerable interest. In this system it was found that the hydrolysis products of *cis*-Pt(II) were more reactive than the parent compound.

A study by SHOOTER *et al.* [17] on the inactivation of bacteriophage has shed very considerable light on the mechanism of action of these Pt compounds. They compared *cis* and *trans* dichlorodiammine platinum(II) and *cis*-dichloroethylenediamine Pt(II) for their ability to inactivate bacteriophages. At the same time they carried out parallel physicochemical studies of the bacteriophages and of the isolated nucleic acids to see if any changes, for example cross-linking of DNA, can be quantitatively correlated with the biological effects. Two bacteriophages were examined, a DNA phage, T7, which contains a single molecule of double-stranded DNA of known molecular weight, and an RNA phage, R17, which contains a molecule of single-stranded RNA. These authors were able to calculate the number of ^{14}C-labelled dichloroethylenediamine Pt(II) molecules bound to either the T7 or R17 phages at the mean lethal doses of the platinum compound, that is, the dose required to reduce survival to 37% and theoretically the dose required to inactivate one phage molecule. It was found that at the mean lethal dose only 1.5 molecules of the platinum compound were bound to each R17 phage and 5 molecules to each T7 phage. Separate physicochemical measurements involving band sedimentation of denatured T7 DNA showed that these platinum compounds could cross-link either isolated phage DNA or the DNA in whole bacteriophage. The cross-linked DNA molecules could be separated from single-stranded DNA molecules due to their different sedimentation characteristics. From the binding studies it was possible to calculate the number of molecules of platinum bound to the DNA when there was on average one cross-link in each DNA molecule. For the ethylenediamine compound it was found that 35 molecules of platinum were bound under these conditions and this shows that cross-linking is a comparatively rare event. Since, as mentioned earlier, only 5 platinum reactions occurred at the mean lethal dose, 4 of which were with the nucleic acid, then it follows that the cross-linking reaction contributes little to the inactivation process. In other words when one has a level of reaction inactivating all phages only very few of them will contain a DNA cross-link, so the cross-link cannot be the basis for the toxic action on the phage.

The ethylenediamine derivative was as effective as *cis*-Pt(II) and both these compounds were appreciably more active than *trans*-Pt(II) in inactivating the bacteriophages. However Shooter *et al.* [17] were able to show that all three compounds cross-linked DNA with approximately equal efficiency and cross-linking was again, in each case, a rare event. This finding confirmed once more the lack of importance of DNA interstrand cross-linking for the inactivation of these bacteriophages. Analogous reasoning to that just described indicated that cross-linking of nucleic acid to protein also contributed little to the inactivation process.

An indication of the likely inactivating event in this particular system can be gained by comparing these results with the platinum compounds with similar experiments using mono- and difunctional alkylating agents. The extent of binding of the ethylenediamine Pt(II) compound and its sensitivity towards the two phages are more comparable to that of the difunctional than they are to that of the monofunctional alkylating agent, so that it seems that platinum compounds do act by reaction

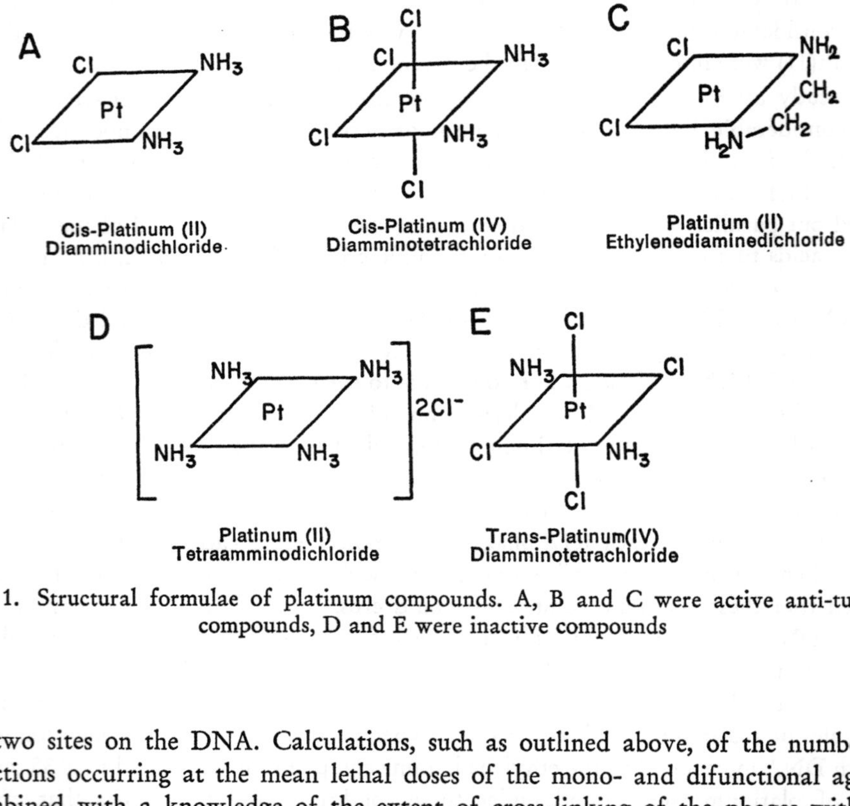

Fig. 1. Structural formulae of platinum compounds. A, B and C were active anti-tumour compounds, D and E were inactive compounds

at two sites on the DNA. Calculations, such as outlined above, of the number of reactions occurring at the mean lethal doses of the mono- and difunctional agents, combined with a knowledge of the extent of cross-linking of the phages with the difunctional agent and of the products of the reaction with DNA indicated that, as with the platinum compounds, interstrand cross-linking of DNA was not likely to be an inactivating event with the difunctional alkylating agents. However, since both single and double-stranded phage molecules are inactivated more readily by difunctional than monofunctional alkylating agents it was suggested that the main contribution to inactivation came from cross-linking of neighbouring bases on the same nucleic acid chain, a reaction which is known to occur with the difunctional alkylating agents. It would therefore appear from these comparative studies between alky-

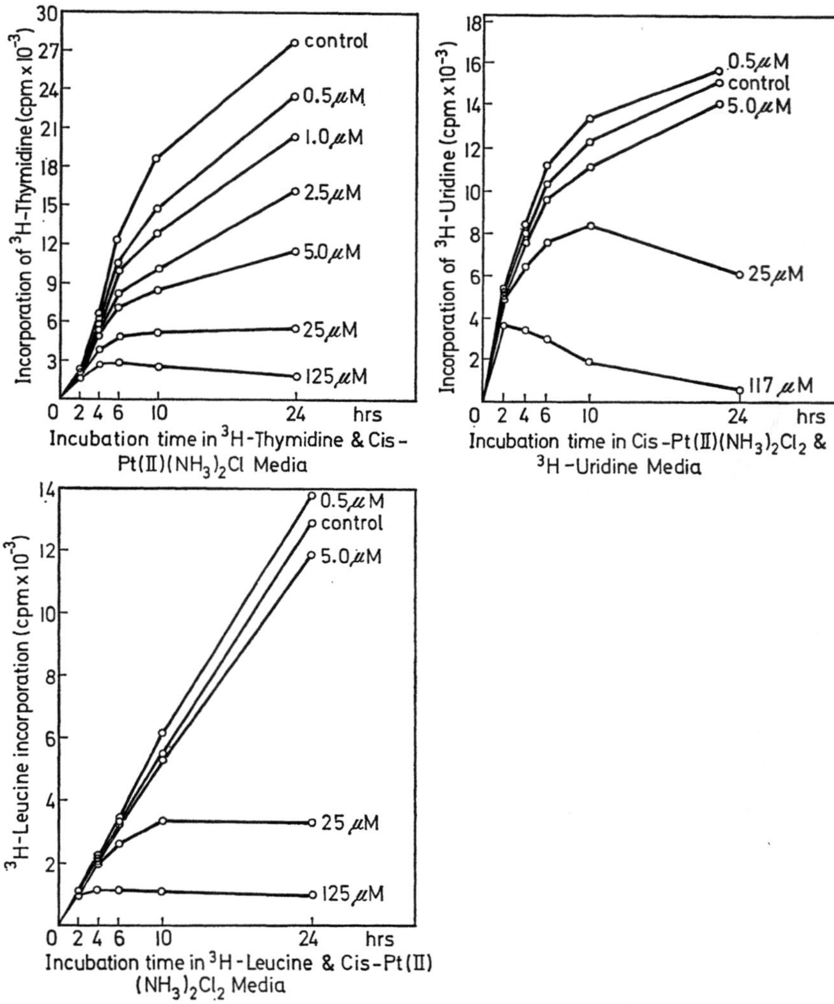

Fig. 2. Effect of *cis*-Pt(II) $(NH_3)_2Cl_2$ on DNA, RNA and protein synthesis in AV_3 cells as measured by the uptake of [^{3}H]thymidine, [^{3}H]uridine and [^{3}H]leucine respectively. Both the platinum compounds and radioactive precursors were applied to cells at time zero

lating agents and platinum compounds that intra-strand cross-linking reactions could be responsible for the inactivation of bacteriophages induced by the platinum(II) compounds.

We can now turn to studies of the effects of these platinum compounds on various biological and biochemical parameters in cells growing in tissue culture.

The first studies in this field were carried out by Harder and Rosenberg [2] and gave a clear indication of the likely mechanism by which mammalian cells were killed by these agents. Previous studies had indicated that those agents which can also inhibit cell division and cause filament formation in some strains of *E. coli*, such as U.V.- or X-irradiation and hydroxyurea, also cause inhibition of DNA synthesis. These authors therefore predicted that platinum compounds would also inhibit DNA

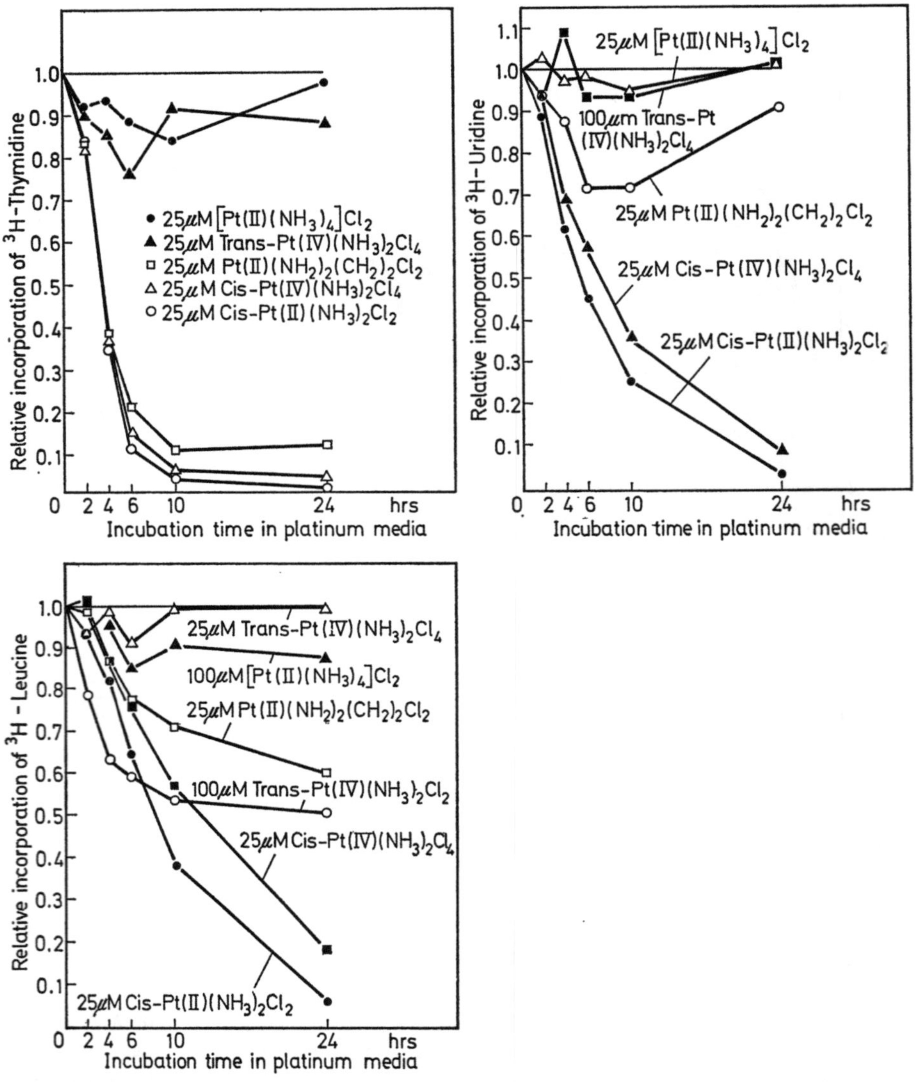

Fig. 3. The relative effects of various active and inactive platinum compounds (see Fig. 1) on DNA, RNA and protein synthesis in AV_3 cells as measured by the uptake of [^{3}H]thymidine, [^{3}H]uridine and [^{3}]leucine respectively. Platinum compounds were applied at time zero and cells were labelled for only the last two hours of treatment at each time period. Results are expressed relative to the control

synthesis in mammalian cells, a biochemical effect which it has been shown results in mitotic arrest of mammalian cells. Harder and Rosenberg compared the effects of three compounds which had been shown to be effective against sarcoma 180 and which also caused filament formation in bacteria (Fig. 1, A, B, C), with two other closely related platinum compounds which did not cause filament formation or inhibit the growth of the sarcoma 180 tumour (Fig. 1, D, E) for their effects on overall DNA,

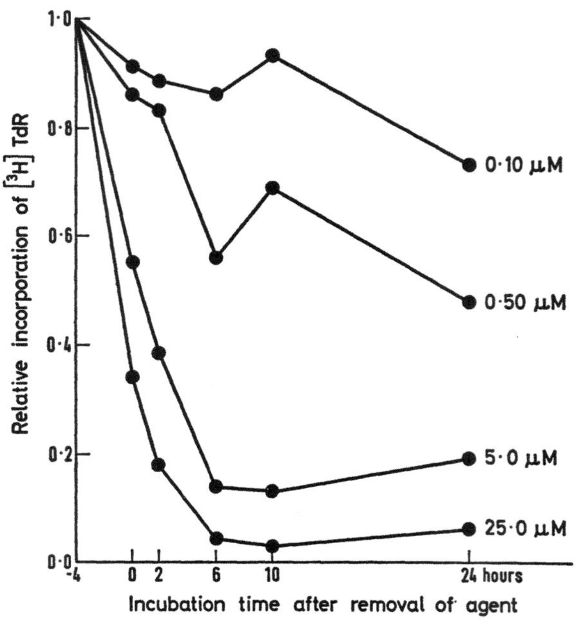

Fig. 4. The relative effect of a 4 hour pretreatment of AV_3 cells with 5 µM *cis*-Pt(II) $(NH_3)_2Cl_2$ on DNA synthesis measured by uptake of $[^3H]$thymidine. Experimental medium was applied at time 4 h and decanted at zero time and the cells washed before applying fresh medium. Cells were labelled for only the last 2 h of incubation for each data point. Results are expressed relative to the control

RNA and protein synthesis. Assays were carried out on human amnion AV3 cells by growing known numbers of cells in plastic petri dishes for various periods of time in the presence of the radioactive precursors, thymidine, uridine and leucine in order to measure the rates of DNA, RNA and protein synthesis respectively. Radioactivity incorporated into trichloracetic acid insoluble material was determined after various periods of time following treatment of the cells with platinum compounds. Fig. 2 compares the effect of various concentrations of *cis*-Pt(II) on DNA, RNA and protein synthesis. It is immediately apparent that whereas a concentration of 5 µM *cis*-Pt(II) has a marked effect on DNA synthesis it has very little effect on RNA or protein synthesis. A concentration of 25 µM *cis*-Pt(II) was considered by these authors to be equivalent to a single injection of 8 mg/kg, the best therapeutic dose in mice for treatment of S-180 assuming a uniform distribution over the entire body. At this concentration DNA synthesis was more selectively inhibited than RNA or protein synthesis although after 24 h all three were virtually completely inhibited. A comparison of the active anti-tumour compounds with the non-active compounds, as seen in Fig. 3, showed that a concentration of 25 µM of the former compounds in every case dramatically inhibited DNA synthesis while the same concentration of the other two showed very little inhibition of DNA synthesis. The active anti-tumour compounds were also able to inhibit RNA and protein synthesis at this concentration but DNA synthesis was most rapidly inhibited followed by that of RNA and protein, as could be seen in Fig. 3. There is therefore a good correlation between the ability of a platinum compound to affect the growth of the sarcoma 180 tumour and its ability to inhibit DNA, RNA and protein synthesis, but notably DNA synthesis in this cell system.

A further significant effect of these agents which these authors described, was the persistent and irreversible nature of the inhibition of DNA synthesis which they produced. If the agent was added for a period of a few hours and then removed, then, as shown in Fig. 4, the rate of DNA synthesis continued to decline during a further period of 10 h. Moreover, this irreversible effect on DNA synthesis is seen even following treatment with very low concentrations of *cis*-Pt(II). In this respect the action of the platinum compounds is quite unlike that of hydroxyurea, which can also inhibit DNA synthesis in mammalian cells, but only while it is present in the cell cultures. The persistent inhibition of DNA synthesis produced by the platinum compound is however reminiscent of that produced by the cytotoxic, anti-tumour, difunctional alkylating agents, such as chlorambucil and sulphur mustard (see below).

HARDER and ROSENBERG [2] were able to show that inhibition of DNA synthesis was not due to any effect of the platinum compounds on the uptake of tritiated thymidine into the acid soluble pool or due to any inhibition of a DNA polymerase. It could therefore be reasonably concluded that the platinum compound bound to DNA and that this was the basis for the preferential effects on DNA synthesis. Higher concentrations of the drugs would be expected to result in more DNA reaction and lead to inhibition of messenger RNA synthesis and this would eventually cause a reduction of protein synthesis. This proposed mechanism of action for the platinum compounds has been shown very convincingly to operate in the case of the difunctional alkylating agents.

Simultaneous studies carried out by HOWLE and GALE [5] are particularly relevant to those of HARDER and ROSENBERG and confirm the selective action on DNA synthesis of these agents. These investigations cannot strictly be classified as tissue culture studies, as these authors used a combined *in vivo-in vitro* system. HOWLE and GALE verified the anti-tumour action of *cis*-Pt(II) against the Ehrlich ascites tumour. The average survival time of treated animals was increased to 37 days compared with 12 days for the saline-injected control animals. DNA, RNA and protein synthesis could be shown to be inhibited in ascites cells which were incubated *in vitro* with the platinum compound, but the greatest effect was on DNA synthesis. However of more significance was their subsequent demonstration of effects on these parameters in cells which had been treated *in vivo*. Animals bearing the ascites tumour were given a single i.p. injection of the drug and at intervals thereafter cells were removed, washed and incubated during a 20′ pulse with either tritiated thymidine, uridine or leucine in order to measure the rates of DNA, RNA and protein synthesis, respectively. Fig. 5 illustrates the result of this experiment and shows two clear effects. Firstly, DNA synthesis was more depressed than either RNA or protein synthesis. Secondly, the effect on DNA synthesis persisted for a considerable period, long after the lesser effects on RNA and protein synthesis were reversed. These findings are therefore very similar to those of HARDER and ROSENBERG [2] using human AV3 cells. The delayed effect on the inhibition of DNA synthesis which was apparent from these independent studies was thought by both groups of workers to indicate that the platinum compound was not necessarily the active agent *per se* but may be converted to the active material. Howle and Gale [4] suggested that possibly more than one metabolic step could be involved. It is of interest therefore to compare some of the effects of the difunctional alkylating agents on DNA, RNA and protein synthesis with those produced by the platinum compounds.

Sulphur mustard has been shown to inhibit DNA synthesis in preference to RNA or protein synthesis in HeLa cells in culture [11]. Moreover, as mentioned earlier, one observes the same temporal effect on DNA synthesis as observed with *cis*-Pt(II), so that the inhibition in rate of DNA synthesis increases to a maximum by 6 h and then persists for many hours thereafter (Fig. 6). With sulphur mustard, however, there can be no possibility of metabolic activation to another species to explain this effect and radioactive labelling studies indicate that the reaction with DNA is complete within a few minutes. Studies using synchronous cultures of HeLa cells showed

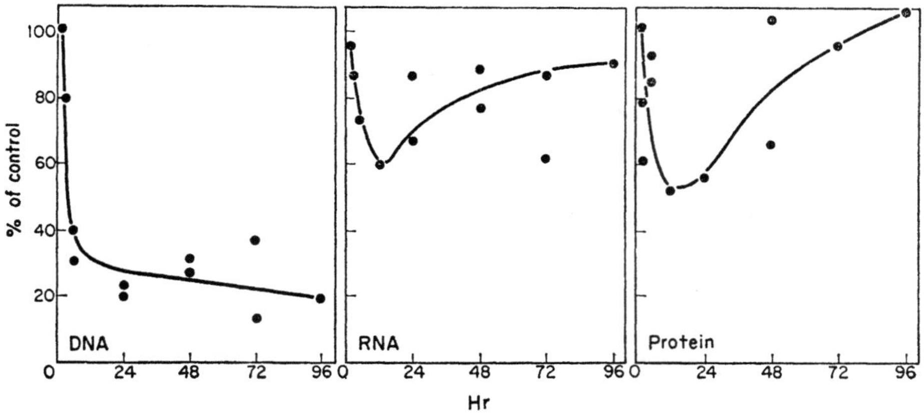

Fig. 5. Effects of a single intraperitoneal injection of *cis*-Pt(II) $(NH_3)_2Cl_2$ at 10 mg/kg in mice on the subsequent synthesis of DNA, RNA and protein by Ehrlich ascites tumour cells when measured *in vitro* at various time intervals after injection

that the apparent increasing effect on the rate of DNA synthesis could be explained, not only by an effect of sulphur mustard on the rate of synthesis of DNA in those cells which were in the S phase at the time of treatment but also by an effect on the progression of cells through the cell cycle. It seems likely therefore that similar changes in cell progression could account for the observations with the platinum compounds; in other words the inhibition to DNA synthesis produces a mitotic block and delays the entry of cells into the next S phase. During this time cells become enlarged and interestingly enough Howle and Gale [4] commented on a similar cell enlargement in their experiments on *cis*-Pt(II) treated ascites cells.

There was a report by Kara *et al.* [6] at the VIIth Chemotherapy Congress in Prague (1971) which confirmed the above effects on DNA synthesis in a rather unusual and interesting manner. Kara had previously shown that when chick embryo fibroblasts were infected with Rous sarcoma virus (RSV) DNA synthesis was stimulated; while Bader has shown that the synthesis of DNA in Rous sarcoma virus-infected cells was necessary not only for the replication of the virus but also it was essential for the malignant transformation of the virally-infected cells. Kara *et al.*

[6] therefore pre-treated their chick embryo cells with *cis*-Pt(II) and then after they were infected with the Rous virus investigated the synthesis of DNA and their transformation. The result clearly showed that *cis*-Pt(II) inhibits irreversibly DNA synthesis in these cells and prevents the induction of cellular DNA synthesis and subsequent malignant transformation after infection with the Rous sarcoma virus. They therefore confirmed BADER's postulate that DNA synthesis in RSV infected cells is essential for the malignant transformation of the infected cells.

I should now like to discuss some of our own published work on the use of tissue culture systems for studying the mechanism of action of the anti-tumour platinum

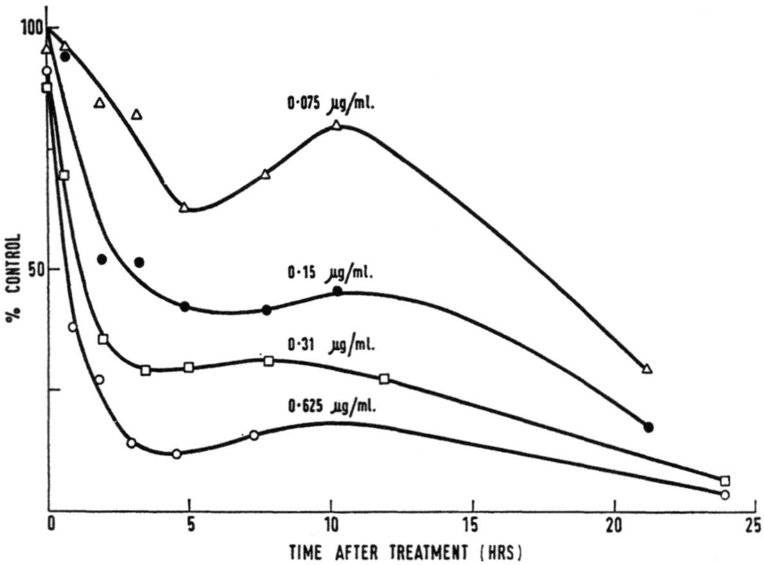

Fig. 6. The effect of mustard gas on DNA synthesis in HeLa cells. Mustard gas was applied to cells at zero time and reaction was complete within 10′. At various times thereafter, at the times shown, the incorporation of [^{3}H]thymidine into DNA during a 20′ pulse was determined and the result expressed relative to the control

compounds. We were concerned to answer three questions which were clearly suggested from the work I have just described on human AV3 cells, and Ehrlich ascites cells and chick embryo cells. The studies of HARDER and ROSENBERG [2] established the correlation between an effect of platinum compounds on animal tumours and an effect on DNA synthesis in cells growing *in vitro*. The first question which we therefore felt it necessary to answer was "is there a correlation between the anti-tumour properties of a compound and its cytotoxic effects on cells growing *in vitro*"? We could then be more confident that the effects we were following *in vitro* such as inhibition of DNA synthesis or reactions with DNA were truly related to the anti-tumour properties of these agents. Secondly, we wished to know if the inhibition of DNA synthesis was causally related to the cell-killing action of the platinum compounds, and thirdly, whether the inhibition of DNA synthesis resulted from the reaction with DNA rather than by inactivation of enzymes involved in DNA synthesis; and more particularly, if there was any evidence that cross-linking of complementary strands of the DNA twin helix by platinum compounds was the critical chemical reaction producing these effects.

Fig. 7 shows the effects of *cis*- and *trans*-Pt(II) on the survival of HeLa cells. Cell survival was measured by the ability of treated cells to divide and form colonies. The *cis* compound is an active anti-tumour compound, while the *trans* isomer is inactive and it can be seen that the *cis* compound is appreciably more toxic to HeLa cells than is the *trans* isomer. Similarly survival curves were constructed for other platinum drugs. We could in this way establish that there is a correlation between the cell-

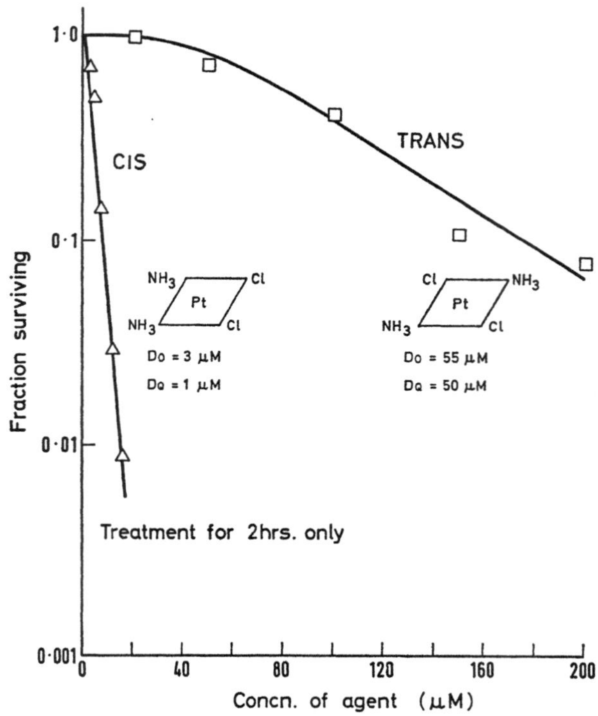

Fig. 7. The effects of *cis*- and *trans*-Pt(II) $(NH_3)_2Cl_2$ on the survival of HeLa cells. Various concentrations of the platinum compounds were applied to HeLa cells in DMSO for a period of 2 h and the effects on colony-forming ability were determined after 14 days

killing effect of these compounds on cells *in vitro* and their anti-tumour action in animals: which means that the difference between the *cis* and *trans* isomers *in vivo* is likely to be due to a basic difference in their mechanism of action rather than to a difference in, say, tissue distribution (Table 2). Inhibition of DNA synthesis could be shown to occur with concentrations of the platinum compounds which resulted in high cell survival, and, in the absence of any other potentially cell-killing biochemical effects. These observations further suggested that inhibition of DNA synthesis was the cause of death in these cells and moreover this inhibition was a consequence of a direct reaction with the DNA (Figs. 8, 9) [13]. The parallel between the toxic effects of the *cis* and *trans* compounds, on the one hand, and the mono- and difunctional

Table 2

	Relative toxicity to HeLa cells [a]	Relative antitumour activity
cis-Pt(II)$(NH_3)_2Cl_2$	1	active
trans-Pt(II)$(NH_3)_2Cl_2$	18.5	inactive
cis-Pt(IV)$(NH_3)_2Cl_4$	0.5	active
trans-Pt(IV)$(NH_3)_2Cl_4$	21.5	inactive

[a] Values obtained from the slopes of the survival curves, D_0, see Fig. 7.

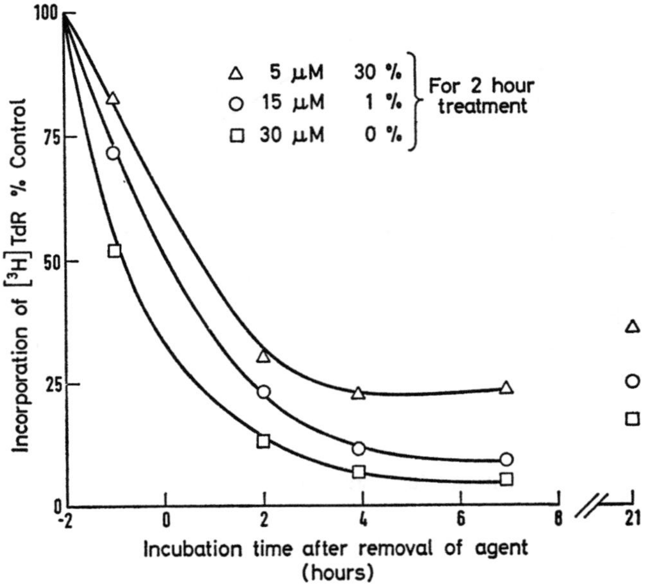

Fig. 8. The effect of *cis*-Pt(II) $(NH_3)_2Cl_2$ on DNA synthesis in HeLa cells. The platinum compound was applied for a period of 2 h. At various times after the addition of the platinum compound aliquots of cells were incubated for 20′ in medium containing [^{3}H]thymidine and the uptake of radioactivity into DNA was determined as a percentage of that in control cells. The effects of a 2 h treatment with three concentrations of the platinum compound on the colony-forming ability of HeLa cells is shown

alkylating agents on the other, pointed to the possibility that the nature of the reaction with DNA was all important in determining these effects on DNA replication. We therefore attempted to find a correlation between one particular possible reaction, namely cross-linking of DNA, and the cytotoxic properties of these compounds. The method we employed has permitted a calculation of the extent of cross-linking at a given level of cell survival (Figs. 10, 11). One strand of DNA was given a density and radioactive label by growing cells in the presence of 5-bromo-2′-deoxyuridine and [^{3}H]TdR simultaneously. Cross-linking between a "heavy labelled" and a "light unlabelled" strand of DNA produced a "labelled hybrid" species of DNA and all these three species could be separated on an alkaline caesium chloride gradient. Both

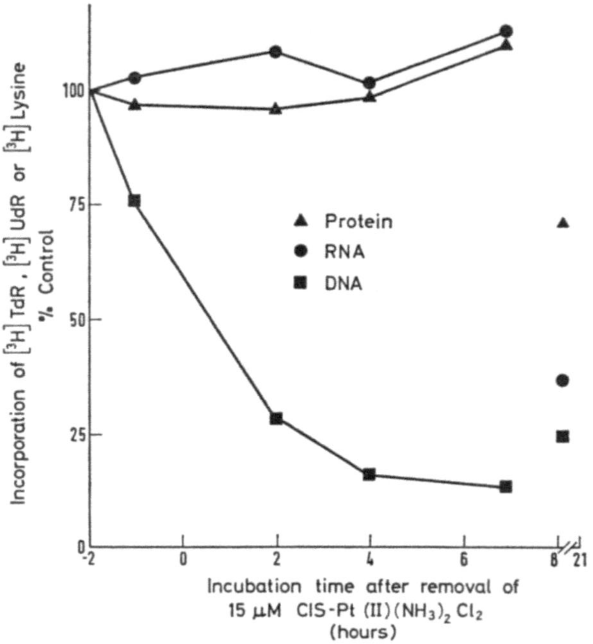

Fig. 9. Relative effects of *cis*-Pt(II) $(NH_3)_2Cl_2$ on DNA, RNA and protein synthesis as measured by the uptake of [^{3}H]thymidine, [^{3}H]uridine and [^{3}H]leucine into HeLa cells

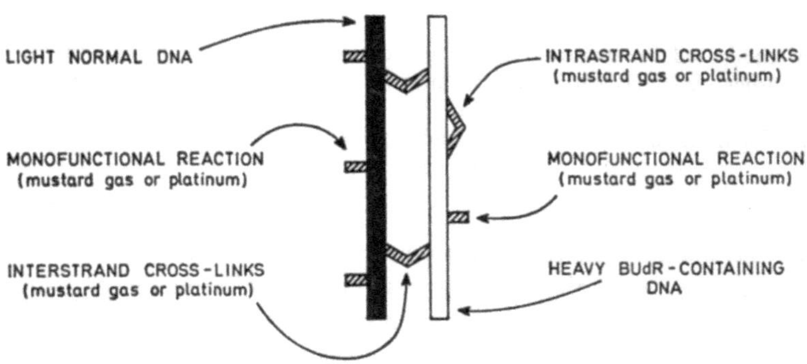

Fig. 10. Schematic representation of reaction of *cis*-Pt(II) $(NH_3)_2Cl_2$ and mustard gas with the DNA of cells which had been grown for one cell doubling time in medium containing 5-bromo-2′-deoxyuridine (BUdR). Several possible reactions with DNA are depicted

mustard gas and *cis*-Pt(II) produced this hybrid species (Fig. 12) [13]. Furthermore we could calculate that very comparable amounts of cross-linking occurred at the D_0 doses for *cis*-Pt(II) and sulphur mustard. There are several potentially reactive groups in DNA including the amino groups of DNA bases. For the *cis*-platinum compound to cross-link, the receptive sites in DNA cannot be greater than 2.8 A apart. A model of DNA shows clearly that amino groups on complementary base pairs are conveniently placed for cross-linking. This is particularly apparent for the 6-amino group of adenine in a ApT sequence, for the 2-amino group of guanine in the narrow

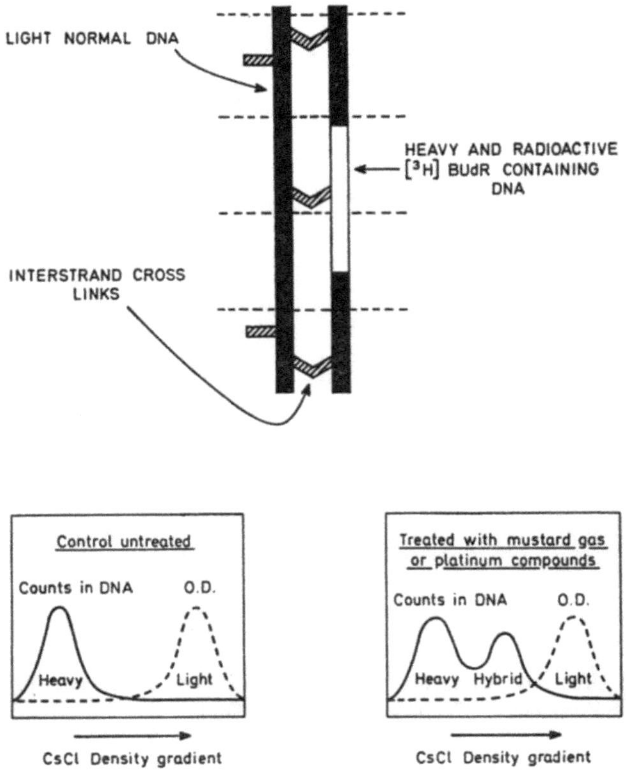

Fig. 11. Method used for quantitating the percentage of cellular DNA which has been cross-linked by either *cis*-Pt(II) $(NH_3)_2Cl_2$ or mustard gas. Cells were grown for a period of 3 h in medium containing [^{3}H]thymidine and BUdR to produce a proportion of labelled hybrid DNA. The fractionation of heavy, light and hybrid DNA was achieved in a caesium chloride isopycnic gradient

groove or the 6-amino groups of cytosine in the wide groove of DNA in a CpG sequence. It seems reasonable to assume that intrastrand cross-linking will also occur although the method described here would not detect such reactions.

These studies have now been extended in order to determine the amount of DNA cross-linking which occurred when cells were treated with a number of platinum compounds at known toxic concentrations. In this manner we hoped to obtain some information on the possible importance of the cross-linking reaction in determining the cytotoxic properties of these agents. These findings will be discussed in some detail later, but in summary it appears that an ability to form DNA inter-strand cross-links could account for the greater toxicity of *cis*-Pt(II) compared with its *trans* isomer. However a similar correlation does not hold when one considers the cross-linking abilities and toxic properties of the *cis*- and *trans*-Pt(IV) compounds. Concomitant studies both on the effects of these various compounds on DNA synthesis and on the levels of reaction which occur with cellular macromolecules nevertheless continue to support the contention that DNA is the essential target for the platinum compound reaction with which results in toxicity in mammalian cells.

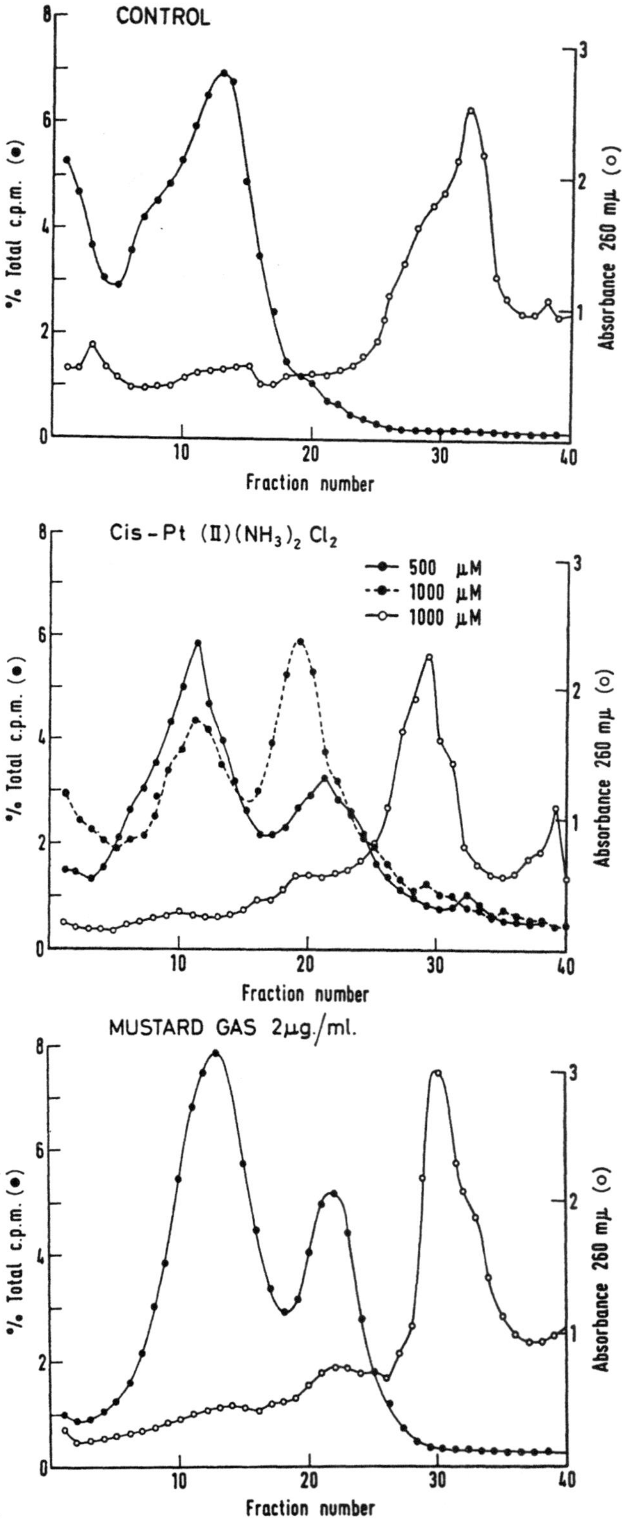

Fig. 12. Formation of cross-linked DNA in HeLa cells with *cis*-Pt(II) $(NH_3)_2Cl_2$ and mustard gas

To summarise briefly, therefore, from the bacterial, viral and tissue culture studies which have been carried out to elucidate the mechanism of action of these new agents, it seems to be fairly clear that reaction with DNA is required to elicit most of the observed biological effects. There is no unequivocal evidence which indicates what is the nature of the essential reaction with DNA although more than one investigation suggests that an ability to react in a bifunctional manner on one strand of DNA, either at two sites on one base or by forming a cross-link between adjacent bases, might be a vital prerequisite for biological activity. Despite many similarities between the properties of platinum compounds and other cytotoxic agents there are indications that the response of some biological systems to these agents are not always identical and this could be related to specific mechanisms involved in the repair of the respective lesions in the DNA. Such differences may hopefully therefore result in these new cytotoxic agents being active against a different spectrum of tumours than the classical DNA reacting alkylating drugs.

References

1. Drobnik, J., Urbánková, M., Krekulová, A.: The effect of *cis*-dichlorodiammine platinum(II) on *Escherichia coli B:* the role of *fil, exr* and *hcr* markers. Mutation Res. **17**, 13 (1973).
2. Harder, H. C., Rosenberg, B.: Inhibitory effects of anti-tumor platinum compounds on DNA, RNA and protein synthesis in mammalian cells *in vitro*. Int. J. Cancer **6**, 207 (1970).
3. Heinemann, B., Howard, A. J.: Induction of lambda bacteriophage in *Escherichia coli* as a screening test for potential anti-tumour agents. Appl. Microbiol. **12**, 234 (1964).
4. Howle, J. A., Gale, G. R.: *Cis*-dichlorodiammine platinum(II) persistent and selective inhibition of deoxyribonucleic acid synthesis *in vivo*. Biochem. Pharmacol. **19**, 2757 (1970).
5. Howle, J. A., Gale, G. R.: *Cis*-dichlorodiammine platinum(II): cytological changes induced in *Escherichia coli*. J. Bact. **103**, 259 (1970).
6. Kara, J., Svoboda, J., Drobnik, J.: *Cis*-dichlorodiammine platinum(II): irreversible inhibition of DNA synthsesis and cell growth in tissue culture and inhibition of chick embryo cell transformation by Rous sarcoma virus. Abstracts of VIIth Int. Chemotherapy Congress. Prague 1971.
7. Kutinova, L., Vonda, V., Drobnik, J.: Inactivation of papovavirus SV40 by *cis*-dichlorodiammine platinum(II). Neoplasma (Bratisl.) **19**, 453 (1972).
8. Renshaw, E., Thomson, A. J.: Tracer studies to locate the site of platinum ions within filamentous and inhibited cells of *Escherichia coli*. J. Bact. **94**, 1915 (1967).
9. Reslova, S., Srogl, M., Drobnik, J.: The transforming activity of *β-subtilis* DNA treated with *cis*-Pt(II) complexes. Abstracts of VIIth Int. Congress of Chemotherapy. Prague 1971.
10. Reslova, S.: Induction of lysogenic strains of *Escherichia coli* by *cis*-dichlorodiammine platinum(II). Chem.-Biol. Interactions **4**, 66 (1971/72).
11. Roberts, J. J., Brent, T. P., Crathorn, A. R.: Evidence for the inactivation and repair of the mammalian DNA template after alkylation by mustard gas and half mustard gas. Europ. J. Cancer **7**, 515 (1971).
12. Roberts, J. J., Pascoe, J. M.: Evidence for cross-linking of complementary strands of DNA in cultured mammalian cells by platinum compounds and mustard gas. Abstracts of VIIth Int. Congress of Chemotherapy. Prague 1971.
13. Roberts, J. J., Pascoe, J. M.: Cross-linking of complementary strands of DNA in mammalian cells by anti-tumour platinum compounds. Nature (Lond.) **235**, 282 (1972).

14. Rosenberg, B., van Camp, L., Krigas, T.: Inhibition of cell division in *Escherichia coli* by electrolysis products from a platinum electrode. Nature (Lond.) **203**, 698 (1965).
15. Rosenberg, B., Renshaw, E., van Camp, L., Hartwick, J., Drobnik, J.: Platinum-induced filamentous growth in *Escherichia coli*. J. Bact. **93**, 716 (1967).
16. Rosenberg, B., van Camp, L., Grimley, E. B., Thomson, A. J.: The inhibition of growth or cell division in *Escherichia coli* by different ionic species of platinum(IV) complexes. J. biol. Chem. **242**, 1347 (1967).
17. Shooter, K. V., Howse, R., Merrifield, R. K., Robins, A. B.: The interaction of platinum(II) compounds with bacteriophages T7 and R17. Chem.-Biol. Interactions **5**, 289 (1972).

Effects of Platinum Compounds on Bacteria, Viruses and Cells in Culture

H. C. Harder (Rapporteur)

1. Aggarwall, S. K., Sodhi, A.: Cytotoxic effects of *cis*-dichlorodiammine-platinum(II) on the mammalian cell *in vitro:* A fine structural study. Dept. of Zoology, Michigan State University, E. Lansing, Michigan 48823.
2. Davidson, J. P., Cleare, M. J., Rosenberg, B.: *In vivo* antiviral activity of *cis*-diaquodiammine platinum(II) against fowl pox virus. Dept. of Pathology and Dept. of Biophysics, Mich. State Univ., E. Lansing, Mich. 48823.
3. Davidson, J. P., Cleare, M. J., Rosenberg, B.: *In vitro* inactivation of Newcastle Disease Virus and Fowl Pox Virus by various platinum coordination complexes. Dept. of Pathology and Dept. of Biophysics, Michigan State University, E. Lansing, Mich. 48823.
4. Drewinko, B., Gottlieb, J. A.: The effect of *cis*-dichlorodiammine-platinum (DDP) on cultured human lymphoma cells. Depts. of Clinical Chemistry and Laboratory of Medicine (BD), and Developmental Therapeutics (J.A.G.), M. D. Anderson Hospital and Tumor Institute, Houston, Texas 77025.
5. Harder, H. C.: Microscopic effects of *cis*-dichlorodiammine platinum(II) on cultured human amnion AV_3 cells, from *Molecular effects of tumor-inhibiting platinum compounds in mammalian cells in vitro: Investigations into their possible mechanisms of action.* (Ph. D. Thesis, Michigan State Univ., 1970). Dept. of Pharmacology, The George Washington Univ., 2300 Eye St., N.W., Washington, D.C. 20037.
6. Pascoe, J. M., Roberts, J. J., Woodward, J.: Interactions between mammalian cell DNA and inorganic platinum(II) compounds. Relationship between DNA interstrand cross-linking and cytotoxic properties. Chester Beatty Research Institute, Fulham Road, London, S.W. 3, and Perkin Elmer Ltd., Maxwell Rd., Beaconsfield, Buckinghamshire.
7. Pascoe, J. N., Roberts, J. J.: Interstrand cross-linking of isolated and cellular DNA by inorganic platinum(IV) compounds. Chester Beatty Research Institute, Fulham Road, London, S.W. 3.
8. Reslova-Vasilukova, S.: Indirect induction of prophage lambda by *cis*-dichlorodiammine platinum(II) treatment. Dept. Biophysics, Faculty of Science, Charles Univ., Vinicra 5. Prague 2, Czechoslovakia.
9. Snejdar, V., Caslavska, J.: Ultrastructural changes in *Escherichia coli* caused by *cis*-dichlorodiammine platinum(II). Laboratory of Electron Microscopy, Dept. of General Microbiology, Institute of Microbiology, Czechoslovak Academy of Sciences. Prague 142-20, Czechoslovakia.

10. TROSKO, J. E.: Mutagenicity *cis*-dichlorodiammine platinum(II) in Chinese hamster cells. Dept. of Human Development, Michigan State Univ., E. Lansing, Mich. 48823.

11. VORISEK, J., SNEJDAR, V.: Study of β-methylgalactoside transport system in *Escherichia coli* as influenced by *cis*-dichlorodiammine platinum(II). Dept. of General Microbiology, Institute of Microbiology, Czechoslovak Academy of Sciences. Prague 142-20, Czechoslovakia.

Originally *cis*-tetrachlorodiammine Pt(IV) was shown to selectively inhibit bacterial cell division, resulting in the formation of *E. coli* filaments in which the ratio of DNA to protein was unchanged. Subsequently, *cis*-dichlorodiammine Pt(II) and other platinum compounds were shown to share this and other radiomimetic properties in both bacterial and mammalian cells. In this section, the ultrastructural, biochemical, antiviral and DNA binding effects of treatment of both bacteria and cultured mammalian cells are examined in an effort to provide additional information which might be useful in elucidating the mechanism(s) of action of these compounds.

Ultrastructural Studies

A. Bacterial Effects

After inducing filaments with an average length of 18 microns by treating a logarithmically growing culture of *E. coli B* with 50 µM *cis*-dichlorodiammine Pt(II) [1], SNEJDAR and CASLAVSKA prepared ultra-thin sections from embedded filaments. Electron micrographic examination revealed that while the cell wall remained intact, the cytoplasmic membrane had extensive contacts with the nuclear material of the cell. The membrane was frequently damaged at these sites (Fig. 1 a). Contact of the nuclear material with the cytoplasmic membrane is an important factor in the coordination of both nuclear and cell division. While this contact is mediated by mesosomes in Gram$^+$ cells, Gram$^-$ bacteria as *E. coli* normally lack such a clearly defined structure, and this contact is maintained by fine structures that are difficult to detect. Because this extensive contact of the nuclear material and cytoplasmic membrane is highly anomalous, it may be directly related to the inhibition of cell division. The nuclear material in the filaments also consistently joined into connecting segments, thereby forming a long axial filament of chromosomal matter. In some filaments the periplasmic space between the cell wall and the cytoplasmic membrane was enlarged such that the cytoplasmic membrane formed envaginations or vacuoles which in some cases were so large that they completely segmented neighboring parts of the cytoplasm. This projection of the cytoplasmic membrane into the cytoplasm is rather unusual in Gram$^-$ bacteria. However, membranes or mesosome-like structures have been observed in the cytoplasm of certain mutant *E. coli* or when normal cell division was impaired.

Ribosomes lost their typical, well defined character and were slightly less numerous. Amorphous inclusion aggregates, presumably ribosomal in nature, also appeared in the cytoplasm. A similar effect on bacterial ribosomes was reported by HOWLE and GALE. Numerous electron transparent granula of stock substances were

---

[1] *cis*-Pt(II) is an abbreviation for *cis*-dichlorodiammine platinum(II).

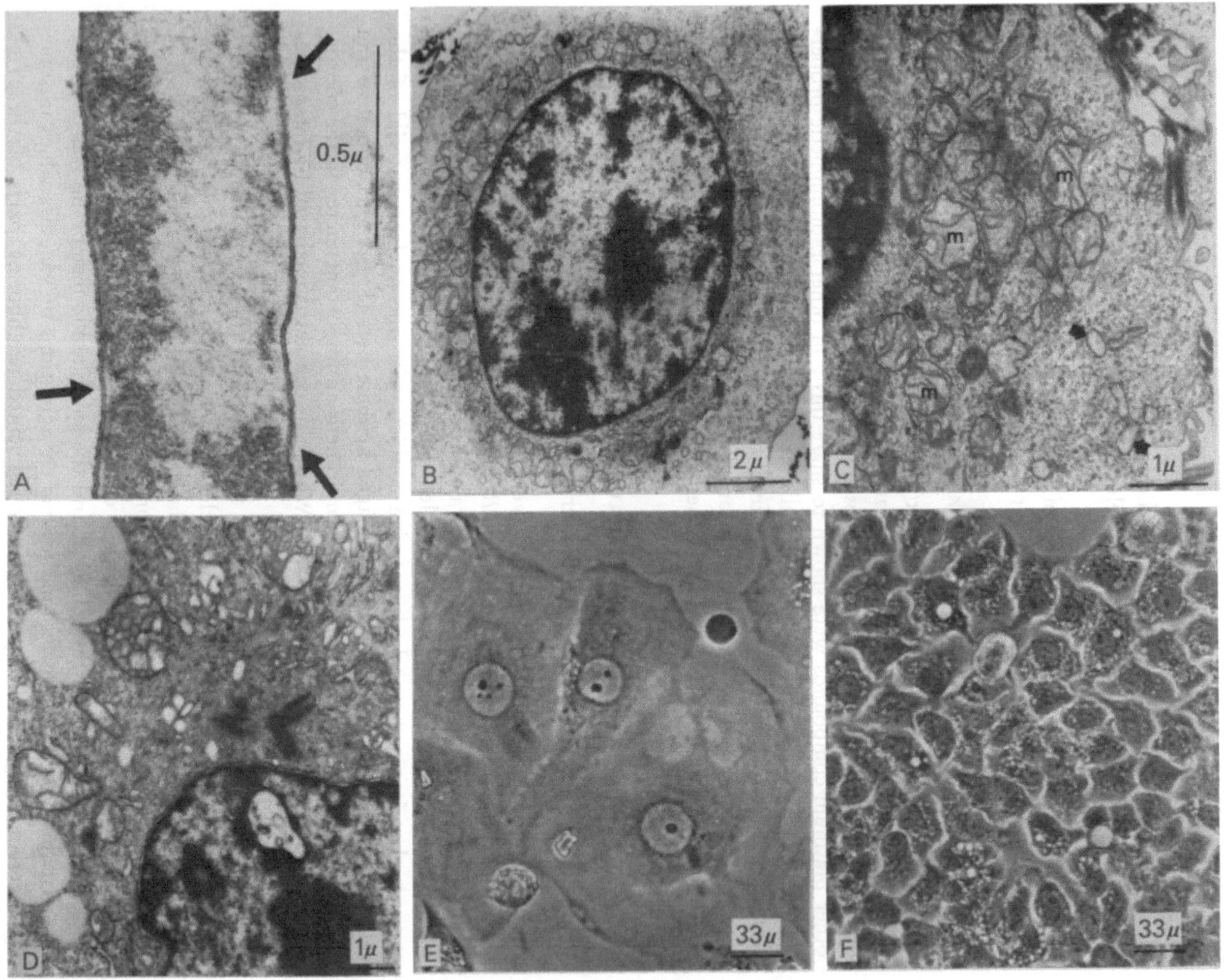

Fig. 1 a to f. Ultrastructural effects of *cis*-dichlorodiammine platinum; a) Electron micrograph of filamentous *E. coli* induced by 3 h of growth with 40 µm *cis*-Pt(II). The nuclear material forms extensive contact with and destroys the cytoplasmic membrane at sites of anomalous contact (indicated by arrows); b) Electron micrograph of Sarcoma-180 ascites cell after 2 h of 5 ppm *cis*-Pt(II) showing the perinuclear band of cellular organelles and clear cortical cytoplasm; c) Higher magnification of part of cell in 1 B showing shrunken, irregular mitochondria (m) and bloated RER cisternae (arrowheads); d) Electron micrograph of Sarcoma-180 cell treated 2 h with 5 ppm *cis*-Pt(II) and then reincubated in drug-free media for 36 h. Note the replicated centiole and some mitochondria with irregular crestae; e) Phase contrast micrograph of giant human amnion AV_3 cells in culture after a 24 h treatment with 5 µm *cis*-Pt(II) and reincubation 96 h in drug-free media; f) Phase contrast micrograph of untreated human amnion AV_3 cells in culture (same magnification as 1 E)

observed at the periphery of the cytoplasm next to the cell membrane. The presence of septa as indications of a constriction process were very rarely observed in the filamentous *E. coli*, indicating that the inhibition of cell division was virtually complete. Where septa were observed, they were derived exclusively from the cytoplasmic membrane; the cell wall did not participate in their formation.

B. Mammalian Cells

In an ultrastructural study performed by AGGARWAL and SODHI, untreated ascites sarcoma-180 cells had many mitotic figures, loosely arranged organelles arranged throughout the cytoplasm with large, oval, dense mitochondria. Cisternae of the rough endoplasmic reticulum were 600—1000 Å apart. Thick, elongated cytoplasmic filaments (60—70 Å) were found throughout the cell but were more concentrated in the perinuclear area and seemed to be associated with the outer nuclear membrane. After *cis*-Pt(II) treatment of the S-180 cells for 15—30 min at 5 ppm (25 μM), all of the microfilaments were concentrated into a discreet band around the nucleus. Golgi cisternae, which were normally concentrated in a juxtanuclear position tended to separate out, and along with the Golgi vessicles, were uniformly distributed in and around the filamentous band. After 60 min of treatment, this band around the nucleus disappeared and morphologically normal mitochondria concentrated around the nucleus, forming a distinct mitochondrial band (Fig. 1 b). The cytoplasm had a decreased electron density while the chromatin appeared granular and more uniformly dispersed. Simultaneously the cisternae of the rough endoplasmic reticulum became distended or bloated. After two hours of treatment these effects became more pronounced and the mitochondria became irregular in shape and very electron lucent (Fig. 1 c).

Following 2 hours of treatment with 5 ppm (25 μM) *cis*-Pt(II), S-180 cells were centrifuged and resuspended in fresh drug-free media to determine if the cells could recover their normal morphology. After 24—36 h of incubation in fresh media, the drug-treated S-180 cells showed a complete recovery of normal morphology. The mitochondria returned to their normal spherical shape, the rough endoplasmic reticulum cisternae were no longer bloated; replication of centriole was common (Fig. 1 d). Cellular organelles in general became evenly distributed throughout cytoplasms. However, no mitotic figures were observed.

In a photomicrographic study of cultured mammalian cells by HARDER, human amnion AV_3 cells were reincubated in drug-free media following a 4—24 h exposure to 5 μM *cis*-Pt(II). Two to four days later, inspection of the cells showed that while cell division had been inhibited completely, cell growth continued, resulting in giant cell formation. In some cases the maximum diameter of these monolayer cells was on the order of 200 microns (Fig. 1 e) in comparison with an average 10—25 micron diameter of untreated cells (Fig. 1 f). Many giant cells appeared quite normal having one giant nuclei while some cells contained 2—6 normal size nuclei. Although the ability of these cells to recover was not quantitatively determined, cytokinesis appeared to occur in some giant cells containing several normal size nuclei 4 days after a 4 h exposure.

Electron micrographs of AV_3 cells by AGGRAWAL and SODHI are similar to S-180 cells up to 2 h of treatment at 5 ppm (25 μM). However, after 4 h of treatment, the rough endoplasmic reticulum cisternae and Golgi membranes became highly vesiculated and the cells started sloughing off huge portions of their cytoplasm, leaving a bare lining of cytoplasm around the nucleus. No mitosis could be observed after 3 days of recovery, although some cells tended to recover normal morphology.

Because centriole replication was observed in S-180 cells 36 h after drug removal in the absence of mitosis figures, AGGRAWAL and SODHI suggested that *cis*-Pt(II) may be inhibiting the synthesis and/or polymerization of spindle fiber proteins in addition to the inhibition of DNA synthesis. Indeed, inhibition of protein and RNA synthesis has been reported in AV_3 cells at 25 μM. While the optimal single therapeutic dose (8 mg/kg) in Swiss white mice is approximately equivalent to 25 μM, the maximum level of drug distributed to solid sarcoma-180 tumors in mice is about 1.4 μg/g tissue (Toth-Allan, Ph. D. thesis), which is approximately equivalent to 5 μM. This suggests that it may be relevant to perform further electron micrographic studies at lower concentrations and in giant mammalian cells as a parallel to the filamentous *E. coli* studies.

Bacterial and Viral Studies

A. Membrane Studies: β-Methylgalactose Transport

Interference with cell membrane or cell wall processes as an alternative hypothesis of the mechanism of bacterial filament formation is suggested by the extensive contact of bacterial nuclear material with the plasma cell membrane in *cis*-Pt(II)-induced filamentous cells and by the fact that penicillin, which inhibits cell wall synthesis, also induces bacterial elongation. The report that *cis*-Pt(II) is bound to the erythrocyte membrane provided the stimulus for VORISEK and SNEJDAR to study possible membrane changes by an investigation of the effects on the galactose active transport system. Their study was confined to β-methylgalactokinase permease system in the galactokinase deficient strain of *E. coli* K 108 in which the 2-methylgalactoside transport system catalyzed only the entry of galactose into the cells. Galactose exit was effected by a distinctly different mechanism.

In untreated cells, ^{14}C-galactose was simply added to a dilute suspension of cells treated with chloramphenicol to stop further growth. Then at various time intervals, samples were filtered through Millipore nitrocellulose membranes and washed prior to measuring the amount of galactose that had accumulated in the cell pool. When 100 μM *cis*-Pt(II) was added simultaneously with the labeled galactose, the uptake of galactose was unaffected. However, with *E. coli* filaments formed after 3 hours of growth in the presence of 40 μM *cis*-Pt(II), the equilibrium level of galactose uptake was about half of the control value when standardized on protein content/ml. A chase-out showed that the initial velocity of the efflux was no greater in the filaments than in the control culture. Hence the lower equilibration value of galactose was most probably due to an inhibition of the uptake system in the filaments. Nevertheless, the lack of an initial effect of *cis*-Pt(II) on uptake when it was added to culture simultaneously with the labeled β-galactose eliminates the possibility of an immediate inhibition of the permease system and/or the inhibition of oxidative phosphorylation supplying energy for the active transport. This absence of a direct effect on galactose transport does not support a mechanism of filament formation based upon the interruption of or interference with active cell membrane processes.

B. *In vivo* and *in vitro* Viral Inactivation

Previous studies in many laboratories have demonstrated the inactivation of many DNA containing bacteriophages including T_{even}, T_3 and ΦX by *cis*-Pt(II). The SV-40 onco virus is also susceptible to inactivation *in vitro* by this drug and its replication is affected in cultured primate cells. Wing web tumors caused by Rous sarcoma virus in chickens can be arrested by *cis*-Pt(II). Following these findings, DAVIDSON, CLEARE and ROSENBERG investigated the inactivation of fowl pox virus (FPV) and Newcastle disease virus (NDV) by the platinum compounds *in vitro* and *in vivo*.

The inactivation of NDV by 600 µg/ml (2 mM) *cis*-Pt(II) *in vitro* proceeded at a slow rate, requiring at least a 6 hour exposure before a drop in the infective titer of the virus particles could be detected (Fig. 2). At longer incubation periods, the inacti-

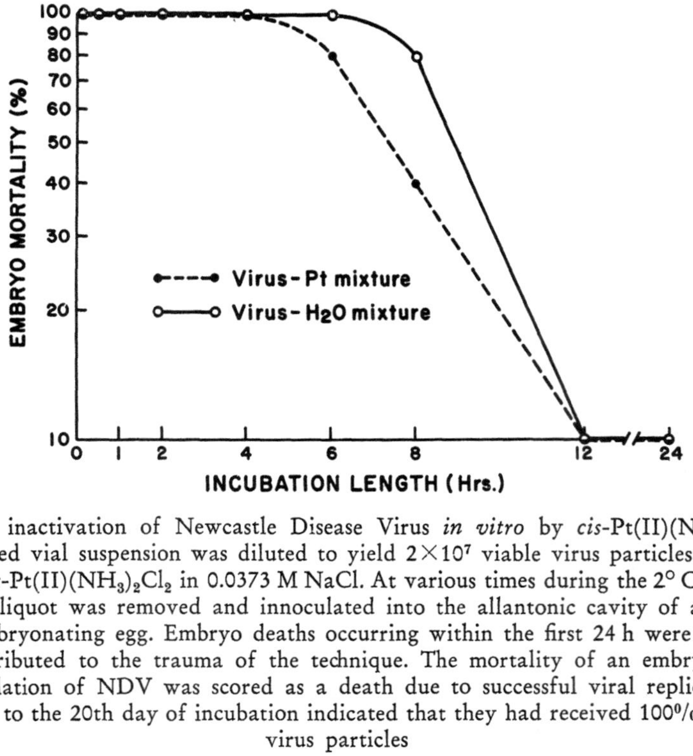

Fig. 2. Viral inactivation of Newcastle Disease Virus *in vitro* by *cis*-$Pt(II)(NH_3)_2Cl_2$. A freshly thawed vial suspension was diluted to yield 2×10^7 viable virus particles per ml and 600 µg/ml *cis*-$Pt(II)(NH_3)_2Cl_2$ in 0.0373 M NaCl. At various times during the 2° C incubation, an 0.20 ml aliquot was removed and innoculated into the allantonic cavity of a previously prepared embryonating egg. Embryo deaths occurring within the first 24 h were disregarded as being attributed to the trauma of the technique. The mortality of an embryo 48—72 h after innoculation of NDV was scored as a death due to successful viral replication. Eggs surviving up to the 20th day of incubation indicated that they had received 100% inactivated virus particles

vation rate of the drug treated viruses soon became identical to the physical (thermal) inactivation of the virus. *Cis*-$Pt(II)(NH_3)_2(H_2O)_2$ however was a potent inactivator of NDV with the infective virus titer being reduced to 40% of the control after an incubation of less than one minute. By 30 min of incubation with this diaquo compound no viable virons could be detected. In contrast, *cis*-$Pt(II)(NH_3)_2C_3H_2O_2$ caused no detectable inactivation of NDV above control levels. As with NDV, when FPV was incubated with *cis*-$Pt(II)(NH_3)_2(H_2O)_2$, inactivation was again so rapid *in vitro* that after an incubation time of one minute, no viable virus particles could be detected.

The results of *in vivo* inactivation of FPV by *cis*-Pt(II)$(NH_3)_2(H_2O)_2$ are given in Table 1 as the average percent reduction of pox lesions relative to the control. The treatment of FPV infected eggs with this drug was highly effective in reducing the number of pox lesions even when the drug was added several hours after virus inoculation. However, other platinum coordination complexes failed to prevent or limit successful virus replication once established within the host. In contrast, none of these platinum compounds, including *cis*-Pt(II)$(NH_3)_2(H_2O)_2$, exhibited *in vivo* antiviral activity against NDV.

Table 1. Anti-viral activity of *cis*-diaquodiammineplatinum(II) against fowl pox virus: *in vivo*

Delay time (hours) between virus & Pt	% Reduction of pox lesion
0	97.2
2	93.3
4	98.2
6	78.2
8	91.3
10	66.7
24	23.7

Fifteen FPV particles suspended in 0.1 ml were innoculated at time zero into an artificial air cell previously prepared on the chorioallantoic membrane (CAM) of each embryonating White Leghorn egg. At various time intervals, 0 or 3.9 μg *cis*-Pt(II)$(NH_3)_2(H_2O)_2$ in 0.2 ml or water was inoculated onto the CAM at the same site which had previously received the virus. On the 5th—6th day of incubation, the CAM was excised, washed and the visible pock lesions in the membrane were counted from treated and untreated eggs.

The *in vitro* measurements indicate that the rate and/or efficiency of viral inactivation is related to the chemical reactivity of the ligands of these compounds. Since the aquo ligands exchange more readily than chloride ligands and chloride ligands more rapidly than the malonate, a reaction with the viron could occur most rapidly with the diaquo platinum compound. This corroborates well with the results of other investigators in demonstrating that *cis*-Pt(II) is most probably active only after the substitution of the chloride ligands for water.

The authors suggested that the complete lack of *in vivo* antiviral activity of the platinum compounds toward NDV may be attributed to a different mode of replication. While both are RNA virons, FPV synthesizes a DNA replicative intermediate whereas NDV replication involves only the synthesis of RNA. Since DNA is selectively inhibited over RNA synthesis in mammalian cells at low concentrations, these results suggest a correlation of *in vivo* antiviral activity with a mode of viral replication involving DNA synthesis. Nevertheless, it remains to be shown whether *in vivo* antiviral activity toward FPV involves an interaction of the platinum with the host cell and if the lack of *in vivo* antiviral activity toward NDV is simply due to the use of an all or nothing bioassay of viral infectivity.

C. Indirect Induction of Prophage

Ultraviolet irradiation of lysogenic bacteria results in a derepression of lytic prophage functions leading to phage reproduction and cell lysis. This process evidently occurs by the appearance of an internal inducer post-irradiation, which, although as yet unidentified, is related to the cell repair mechanism. The irradiation target triggering this induction process is the bacterial chromosome or other chromosome-like replicons, e. g., a sex-specific F factor. With direct induction, the inducer, produced following irradiation of the cell, inactivates the phage repressor in that cell leading to the lytic response. In contrast, indirect induction can occur when the F factor is transferred from an irradiated nonlysogenic donor into a lysogenic, non-

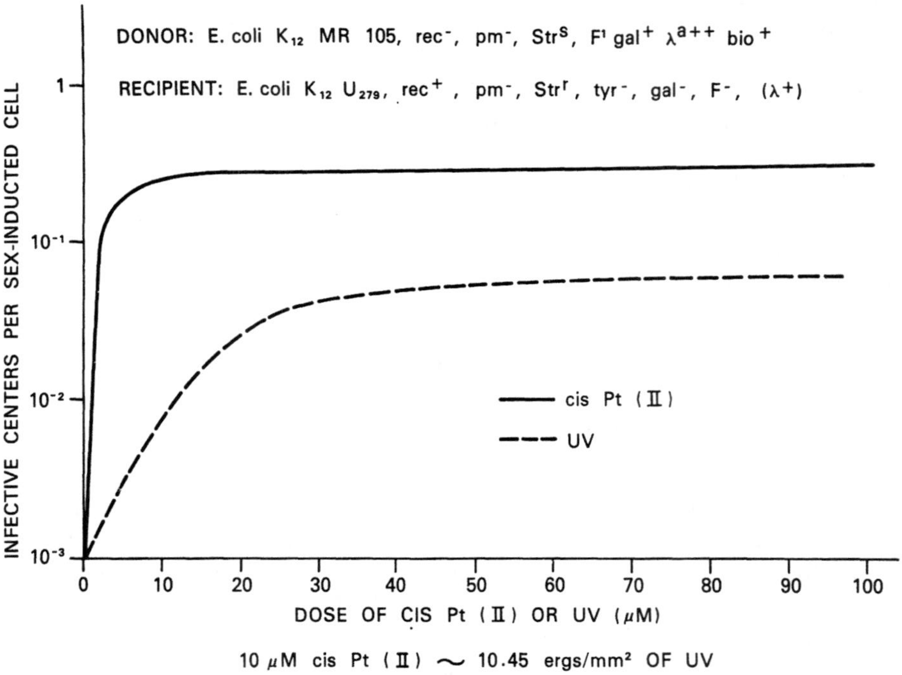

Fig. 3. Indirect prophage induction by ultraviolet light and *cis*-$Pt(II)(NH_3)_2(H_2O)_2$. The donor F′ gal$^+$ cells were either irradiated with UV or treated 10 min with *cis*-Pt(II) aquated by previous incubation for 1 h at 71° C. To test for survival, a portion of the treated cells were plated for colony formation ability (CFA). Another aliquot of treated cells were mated with F$^-$ gal$^-$ Str^r and plated for gal$^+$ colonies on selective media to test for the efficiency of F$^-$ gal$^+$ transfer. A third fraction of treated cells were mated F$^-$ gal$^-$ (λ) recipients and plated for infective centers to test for indirect induction

irradiated recipient by conjugation. In this case, prophage induction occurs even if the lysogenized recipient cells have not been exposed to the inducing agent. Since DNA is the sole component which enters into the recipient cell by conjugation, the phenomenon of indirect induction proves the damaged DNA is the only target factor responsible for phage induction.

Like radiation, *cis*-Pt(II) is also an efficient and highly specific phage inducing agent which acts directly on lysogenic cells. Since it is known that *cis*-Pt(II) reacts readily with RNA and protein as well as DNA both *in vivo* and *in vitro*, Reslova-Vasilukova decided to determine if *cis*-Pt(II)-DNA lesions are sufficient to cause prophage induction by employing the indirect prophage induction method described above. The fraction of cells receiving the F′ gal$^+$ factor that were induced (indirect sexductants) was higher with aquated *cis*-Pt(II) than by UV. This is shown in Fig. 3 in which the UV dose and aquated *cis*-Pt(II) concentrations are given on a scale of equivalent lethal effect. The LD_{50} dose of aquated *cis*-Pt(II), 90 µM for a 10 min exposure, enhanced the infective center (IC) by 290 compared with an enhancement of 60 at the LD_{50} dose of UV, 94 ergs/mm^2. Premature interruption of conjugation before 20 minutes prevented the enhanced formation of infective centers, while the transfer of the gal$^+$ marker was only slightly affected when mating was interrupted after only 4 min. This showed that essentially the entire F factor must be transferred to obtain enhanced indirect induction. Furthermore, because near maximum phage induction occurred at very low concentrations (2—4 µM) which do not significantly affect cell survival, *cis*-Pt(II) is a more specific inducer than UV. Since only DNA bound platinum could enter the recipient cells, free *cis*-Pt(II) could not have caused this induction. Therefore the author concluded that the reaction of *cis*-Pt(II) with the chromosome or chromosome-like replicon of DNA is a sufficient lesion and the only site of attack that could lead to indirect induction, independent of the nature of the chemical damage.

Mammalian Cell Studies

A. Cell Cycle Survival Studies

In a study to evaluate the effect of *cis*-Pt(II) on the viability of a human lymphoma cell line (T_1) in culture, Drewinko and Gottlieb demonstrated that the mean lethal dose for asynchronous cell survival (D_0) was approximately 5.5 µg/ml (18 µM) for a one-hour treatment. By comparison with adriamycin, methotrexate, cytosine arabinoside, BCNU and bleomycin, *cis*-Pt(II) was a highly effective chemotherapeutic agent. Only adriamycin appeared more cytotoxic on a mg/ml basis.

To determine the stage in the cell cycle that is most sensitive, T_1 cells were synchronized by thymidine block and mitotic selection after release from a thymidine block. In all experiments synchrony as monitored by the labeling index was found to be 90% for cells in S phase, 70% to 80% in G_2 and 70% in G_1. The survival of T_1 cells treated with 5 or 10 µg/ml *cis*-Pt(II) for 1 h during various stages of the cell cycle is shown in Fig. 4. While there was some fluctuation with G_1 and G_2 cells appearing more sensitive than mid S phase cells, the difference in survival of the most and least sensitive was on the order of 5-fold at 10 µg/ml and 2-fold at 5 µg/ml. This small difference in phase sensitivity was further emphasized in survival curves of synchronized cells treated over a broad range of concentrations for 1 h. From such curves, the survival D_0's were calculated to be 4, 6 and 5 µg/ml for cells in G_1, mid S and G_2 phases respeectively. Since the cytotoxicity of this drug was similar in all phases, except possibly M, which was not tested, it was concluded that *cis*-Pt(II) is not a phase-specific drug.

If the mechanism of *cis*-Pt(II) involved the inactivation or inhibition of the synthesis of certain DNA precursors or mitotic enzymes, selective toxicity would be expected during that phase in which the activity of those enzymes is essential. The lack of such phase-specific cytotoxicity with *cis*-Pt(II) mitigates against an enzyme inactivation hypothesis. However, these results are consistent with other radiomimetic properties of this drug.

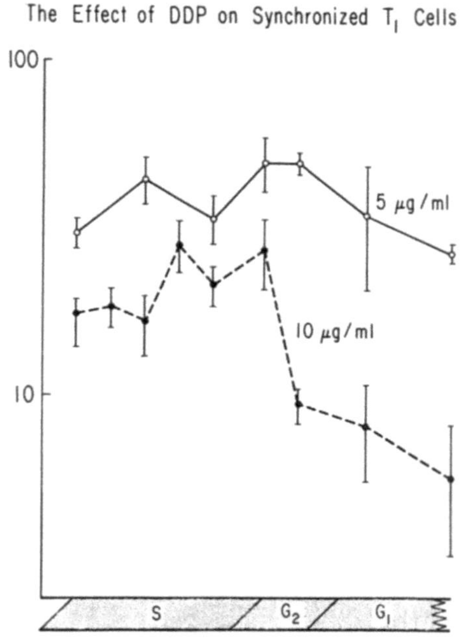

Fig. 4. The effect of *cis*-Pt(II)$(NH_3)_2Cl_2$ synchronized T_1 cells

Table 2. Effect of *cis*-Pt(II)$(NH_3)_2Cl_2$ on the frequency of mutations from azaguanine sensitivity to resistance in chinese hamster cells in culture

Experiment	Treatment	Survival %	No. mutant colonies per plate	Total No. of viable cells ($\times 10^5$)	Mutation frequency per 10^5 survivors
1	Control	100	24	15	1.60
	5 µm *cis*-Pt(II) $(NH_3)_2Cl_2 \times 4$ h	33	18	4.95	3.63
2	Control	100	56	30	1.86
	5 µm *cis*-Pt(II) $(NH_3)_2Cl_2 \times 4$ h	22	67	6.3	10.68

B. Mutagenicity

To test the mutagenicity of *cis*-Pt(II), Trosko employed the quantitative method of Chu and Malling for the detection of forward mutations to 8-azaguanine resistance in Chinese hamster cells (V 79-4). Following the exposure of 1.5×10^5 cells on plates to fresh 5 µm *cis*-Pt(II) in complete media for 4 h the media was decanted and the cells were rinsed and reincubated with fresh media for another 48 h. Then to test for forward mutations, 8-azaguanine was added to both treated and untreated plates which were scored two weeks later for surviving mutant colonies. The data shown in Table 2 indicates that a 4 h exposure to 5 µm *cis*-Pt(II) was both highly toxic and mutagenic. Mutagenicity of *cis*-Pt(II) is consistent with the other radiomimetic properties of this drug and is additional evidence that the mechanism of action involves a reaction of platinum with DNA.

C. *In vivo* and *in vitro* DNA Crosslinking Studies of Platinum Compounds in Relationship to the Cytotoxicity

In a study of the toxic, macromolecular binding and DNA cross-linking effects of platinum II compounds, Pascoe, Roberts and Woodward initially demonstrated the effects of *cis*- and *trans*-dichlorodiammine Pt(II) on the survival of HeLa cells following a 2 h treatment. As shown in Fig. 5, the active antitumor agent, *cis*-Pt(II) was much more toxic than the corresponding inactive *trans* isomer. The ratio of the slopes (i. e., the D_0 values for exposure) of the *trans*- and *cis*-Pt(II) curves was 18. The parallel between the antitumor activity and HeLa cell toxicity indicated that this was a valid model system to pursue further mechanistic studies to determine the cause of their different cytotoxicities.

First they investigated the possibility that the lower ability of the *trans* isomer to kill cells might be due to its inability to enter cells and react with DNA, RNA or protein. Fig. 6 shows the results of such binding studies at low concentrations where there was cell kill only with the *cis* isomer. Noting the different scales, one can see that approximately twice as much of the *trans* isomer was bound to these macromolecules as the *cis* isomer. At higher concentrations where significant cell kill also occurred with the *trans* isomer, the binding of *cis*- and *trans*-Pt(II) to DNA and RNA was of the same order of magnitude. Hence the *trans* isomer does not owe its low cytotoxicity to its failure to react with cellular macromolecules.

To assess the possible importance of DNA, RNA and protein as primary targets for platinum II compounds, these binding data were used to construct curves of log cell survival against the amount of drug bound to each type of macromolecule. The resulting graphs were then characterized in a similar way to a curve of log cell survival *versus* the dose of drug given to the cells. The shoulder width of the binding curve was given by B_q and the slope of the straight line portion by B_0. The binding coefficients, B_q and B_0, for the binding of each isomer to DNA, RNA and protein are given in Table 3. For both *cis*- and *trans*-Pt(II), the binding coefficients were higher for RNA than DNA and the values for protein binding in the case of the *cis* isomer were lower than those for DNA. The significance of the B_0 values became most apparent when they were used to calculate the number of platinum molecules bound to each macromolecule. Table 4 shows the approximate molecular weights of

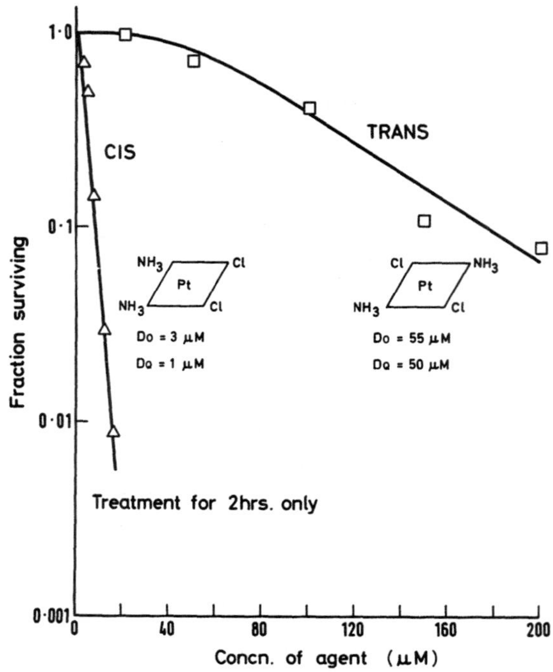

Fig. 5. Effects of *cis*- and *trans*-Pt(II)$(NH_3)_2Cl_2$ on the survival of HeLa cells

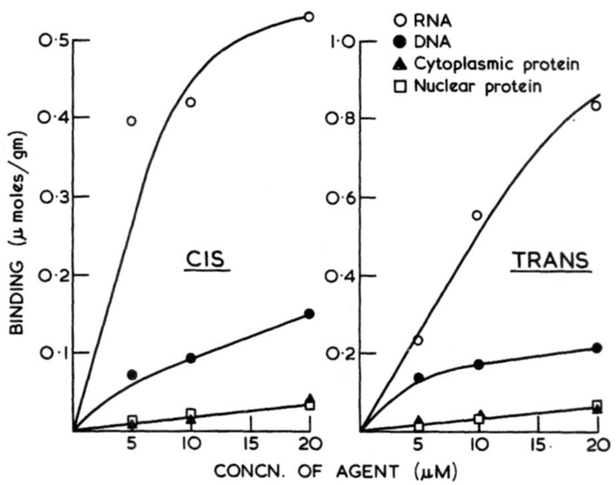

Fig. 6. Binding of *cis*- and *trans*-Pt(II)$(NH_3)_2Cl_2$ to HeLa cell macromolecules

DNA, RNA and protein, and the results of such calculations performed at the concentration which reduced the surviving cell fraction from f to 0.37 f. Because there are more platinum molecules bound per DNA molecule than per RNA or protein molecule, these calculations would suggest that DNA is the most sensitive cellular target for both *cis*- and *trans*-dichlorediammine Pt(II).

Table 3. Binding coefficients of *cis*- and *trans*-$Pt(II)(NH_3)_2Cl_2$ to macromolecules in HeLa cells

Compound	Binding to macromolecules					
	B_q (μmoles/gm) (shoulder width)			B_0 (μmoles/gm) (slope of curve)		
	DNA	RNA	Protein	DNA	RNA	Protein
cis-$Pt(II)(NH_3)_2Cl_2$	0.045	0.300	0.002	0.0225	0.030	0.00675
trans-$Pt(II)(NH_3)_2Cl_2$	0.170	0.300	—	0.125	0.650	—

Table 4. Yield of *cis*- or *trans*-$Pt(II)(NH_3)_2Cl_2$ bound to HeLa cell macromolecules of average molecular weight at the mean lethal dose

Molecule	Approx. mol. wt.	*cis*-$Pt(II)(NH_3)_2Cl_2$		*trans*-$Pt(II)(NH_3)_2Cl_2$	
		B_0 μmoles/gm	No. Pt mol. bound when surviving fraction reduced from f to 0.37 f	B_0 μmoles/gm	No. Pt mol. bound when surviving fraction reduced from f to 0.37 f
DNA	1×10^9	0.0225	22 Pt/DNA mol.	0.125	125 Pt/DNA mol.
mRNA	4×10^6		1 Pt/8 mRNA mol.		2.5 Pt/mRNA mol.
rRNA	$0.5-1\times10^6$	0.030	1 Pt/30 rRNA mol.	0.650	1 Pt/2 rRNA mol.
tRNA	2.5×10^4		1 Pt/1500 tRNA mol.		1 Pt/70 tRNA mol.
Protein	1×10^5	0.00675	1 Pt/1500 protein mol.	—	—

DNA interstrand cross-linking has been suggested as an important cytotoxic event in the case of the bifunctional alkylating agents. That this is also important in the case of the antitumor platinum compounds was indicated by the fact that a Walker carcinoma with an acquired mephalan resistance is cross-resistant to *cis*-Pt(II). To detect DNA interstrand cross-linking, HeLa cells were grown for a short period in radioactively labeled bromodeoxyuridine. Alkaline cesium chloride density gradient centrifugation of DNA isolated from untreated cells resulted in a gradient having two DNA peaks, one heavy radioactive DNA peak and one light non-radioactive DNA peak measured optically. DNA containing interstrand cross-links gave 3 peaks, heavy radioactive DNA, hybrid-density, radioactive DNA (cross-linked material) and non-radioactive light DNA. From the areas under the hybrid, heavy and light DNA peaks, the percent of cross-linked DNA was calculated. Fig. 7 shows that in the straight portions of the curves, there is on the order of a 12-fold difference in the ability of *cis*- and *trans*-Pt(II) to cross-link the DNA of whole cells. Comparison of this value with the relative ability of these compounds to kill cells measured either on the basis of exposure dose ($D_0^{trans}/D_0^{cis} = 18$) or DNA binding dose ($B_0^{trans}/B_0^{cis} = 5.6$) needed to produce an equal effect on cell survival would indicate that the ability of *cis*- and *trans*-dichlorodiammine Pt(II) to cross-link cellular DNA could be related to their cytotoxic properties.

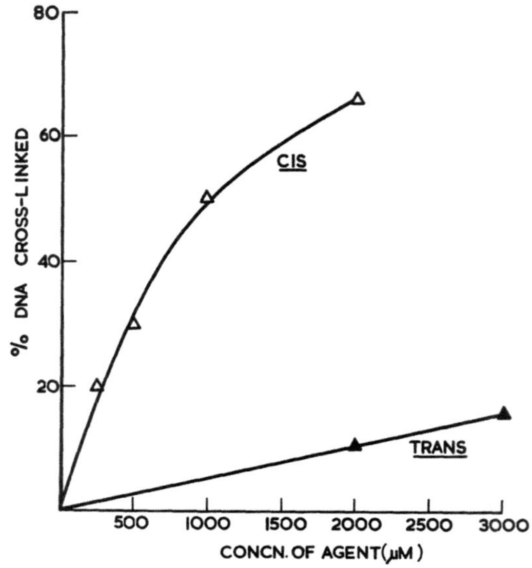

Fig. 7. Cross-linking of HeLa cell DNA *in vivo* by *cis*- and *trans*-$Pt(II)(NH_3)_2Cl_2$

Table 5. The dose of *cis*- and *trans*-Pt(II) and (IV) compounds required to cross-link HeLa cell DNA *in vivo* and *in vitro*

Compound	Dose required to produce 10% DNA cross-linking *in vitro* (A) μm	Dose required to produce 10% DNA cross-linking *in vivo* (B) μm	Ratio of A : B
cis-$Pt(II)(NH_3)_2Cl_2$	0.5	150	1 : 300
trans-$Pt(II)(NH_3)_2Cl_2$	1.0	1900	1 : 1900
cis-$Pt(IV)(NH_3)_2Cl_4$	33.0	420	1 : 12.7
trans-$Pt(IV)(NH_3)_2Cl_4$	67.0	570	1 : 8.5

From the amount of DNA binding and cross-linking in whole cells, the frequency of DNA interstrand cross-links introduced into cellular DNA was calculated. With the *cis*-Pt(II), only 1 in 400 DNA reactions resulted in a cross-link, whereas the value for the *trans* isomer was only 1 in 4000 reactions. In contrast. 1 in 8 reactions of sulfur mustard with DNA *in vivo* is an interstrand cross-link. Hence, the far less frequent formation of DNA interstrand cross-links per DNA reaction could indicate that this is unlikely to be a major factor in causing cell death. However, the number of DNA interstrand cross-links which must be introduced into HeLa cells to reduce the surviving fraction from an f to 0.37 f is of the same order, being approximately 350 for the *cis* isomer and 200 for the *trans* isomer. Therefore, if DNA interstrand cross-linking is not the primary cell killing mechanism for *cis*- and *trans*-Pt(II), it appears at least to parallel the major cytotoxic lesion. The primary cytotoxic lesion could well be another type of cross-link such as a DNA intrastrand cross-link or one between DNA and proteins.

In a second paper, PASCOE and ROBERTS performed analogous cross-linking studies with *cis*- and *trans*-tetrachlorodiammine Pt(IV). The slopes of the HeLa cell survival versus drug concentration curves for a 2 h exposure were 1.5 and 65 μm respectively. Hence the tumor inhibiting *cis* isomer is more than forty times as toxic as the inactive *trans*-Pt(IV) compound. Using alkaline cesium chloride density centrifugation they then performed *in vitro* cross-linking studies by treating HeLa DNA after isolation from cells pulse-labeled with radioactive bromodeoxyuridine. With these drugs, as shown in Table 5, there was only a 2-fold difference in their ability to produce 10% cross-linking *in vitro*. A similar small difference in *in vitro* cross-linking ability was observed between the *cis*- and *trans*-dichlorodiammine Pt(II). However, the platinum(IV) compounds were far less reactive than their platinum(II) analogues because at least 50 times as much platinum(IV) compounds were required to produce the same extent of cross-linking as measured with the platinum(II) compounds *in vitro*.

When cultured HeLa cells were treated prior to isolation of the DNA, the difference in the cross-linking ability by *cis*- and *trans*-Pt(IV) was still of the order of 1.5-fold, as shown in Table 5. In contrast, the difference in cross-linking ability by *cis*- and *trans*-Pt(II) was much larger, being of the order of 12. Considering the very small difference in cross-linking ability and the large difference in the cytotoxicity, it would appear that, in contrast to the Pt(II) analogues, the cross-linking of HeLa cell DNA *in vivo* by *cis*- and *trans*-Pt(IV) tetrachlorodiammine has little or no relationship to their cytotoxic properties.

Animal Studies

Anti-Tumour Effects of Platinum Complexes in Experimental Animals

T. A. Connors

The discovery of new types of anti-cancer agent is not always the result of large scale screening or of working hypotheses which suggest that a particular chemical may have anti-tumour properties. Quite frequently new agents are found because of unexpected observations which may be made in fields completely unrelated to cancer chemotherapy.

The discovery of the platinum complexes with anti-tumour activity is one such example of how a chance finding followed up by elegant research opens up the way to the investigation of a new chemical class of chemotherapeutic agent [17]. Extensive testing against laboratory models is then required to determine the potential value of the agent in the clinical treatment of cancer. There are three preliminary objectives in the study of a new class of anti-cancer agent.

Firstly, it must be known whether the chemical represents a completely new type of agent in its biological properties. It must be established whether or not its mechanism of action is similar to any drug of established clinical use. This is of particular importance since clinical experience has confirmed what was clear from animal experiments, that no chemical is effective against more than a small number of cancer types. Although good responses may be consistently obtained against some neoplasms, other cancers are invariably resistant. A new class of agent acting by completely different mechanisms may conceivably be effective against those tumours, particularly the solid carcinomas, which do not at present respond to any form of chemotherapy. The platinum complexes have therefore been tested against a wide range of animal tumours and their spectrum of action compared with clinically useful agents to determine whether important differences exist. Many tumour lines are employed consisting mainly of transplanted tumours but also chemically and virally induced tumours. Past experience has shown that some tumours are more reliable than others in predicting clinical value and that agents with a wide spectrum of action, i. e. effective against a range of tumours, are more likely to be clinically useful than those with good anti-tumour action but limited to only one or two tumour lines. Further testing on tumours with acquired resistance to another class of agent enables the mechanism of action of the new agent to be further evaluated. A cross resistance indicates that somewhere in their mechanism of action the two classes of compound share a common pathway.

Secondly, it is of importance to determine whether the new class of agent is synergistic with any clinically useful compound. Even the best anti-cancer agents are not entirely selective in their action against malignant cells, and at effective dose levels invariably damage other tissues so that, in the majority of cases, toxicity prevents treatment being continued until every cancer cell has been eradicated. Two agents which are effective against the same tumour cell type but which differ in their major side effects may be administered in combination so that the tumour cell kill is additive but the toxicity is not. In such cases of 'therapeutic synergism' ([22]) a greater tumour cell kill can be achieved with a two drug combination than can be obtained with either drug alone, even at their maximum tolerated doses. Combinations of as many as eight drugs have been used clinically on an empirical basis. To test drug combinations adequately in laboratory animals is difficult because of the permutations of drugs, dose schedules and sequences of administration. However, simple two drug combinations tested against animal tumours do give an indication as to whether or not a particular combination shows therapeutic synergism and can exclude the possibility that they are antagonistic to one another.

Thirdly, analogues of the active new agent are synthesised and tested in the search for derivatives which may have fewer side effects and greater selectivity. Often the first member of a series to have a clinical trial becomes the preferred drug. This is because analogues synthesised at a later date, possibly when considerable clinical experience has been gained with the parent compound, have to be comparable or considerably better in order to compete with the original drug. Methotrexate, for example, was one of the first anti-folates to be synthesised and it remains the drug of choice despite the discovery of many more folate antagonists with good activity against experimental tumours. In some cases continued synthesis of chemicals related to a clinically useful one does uncover compounds of greater selectivity. Cyclophosphamide was first made about nine years after the clinical trials of nitrogen mustard (HN2) and it is now one of the most extensively used chemotherapeutic agents. The rational synthesis and testing of a limited number of analogues of an active compound is in any case essential in studies on mechanism of action. Definition of the structural requirements of a molecule for anti-tumour effectiveness and comparison of the properties of active and inactive members of the series is usually the first step in understanding how a chemical acts as an anti-cancer agent. Synthesis of analogues is also important if one wishes to improve the ease with which a compound may be administered. Some agents may have good anti-tumour properties but be difficult to handle because they may be, for instance, unstable, vesicant or insoluble. The design of active analogues without these properties can greatly assist their large scale clinical usage.

Testing Against Animal Tumours

Early attempts to discover anti-cancer agents were based on the assumptions that cancer at the biochemical level was one disease and not a group of related but different diseases, and that animal tumours were predictable models for human cancer. After thirty years of cancer chemotherapy it is now quite clear that each class and even sub-class of cancer responds differently to a particular drug and that no animal model exists which is completely reliable in selecting agents for any type of human cancer. Screening tests nowadays utilise a variety of models and evaluation of a new agent is made from its anti-tumour selectivity, its spectrum of action in a range of tests, its properties compared with known agents and from its preclinical pharmacology. In its clinical trial the poor predictability of animal systems is recognised by testing against as many types of human cancer as possible.

Table 1 gives examples of the types of test that may be used in the search for new anti-cancer agents. *In vitro* systems using mammalian, bacterial and other cell types in culture are essentially tests to detect agents which inhibit cell division, and a compound passing as active is one that can prevent growth at low concentrations. The platinum complexes such as *cis*-dichlorodiammine platinum(II) (*cis*-Pt(II)) are readily selected as active since they can completely inhibit the growth of many types of cell

Table 1. Types of test system used to detect anti-cancer agents

in vitro tests	*in vivo–in vitro* tests
Biochemical assays	Bioassays
Plant and bacterial cells	Colony forming assays
Mammalian cancer cells	
Cell or organ culture	
in vivo	Heterotransplanted human tumours
Spontaneous tumours	Hamster cheek pouch
Chemically or virally induced tumours	Conditioned animals
Transplanted tumours	Chorioallantoic membrane

at concentrations of 1 μg/ml or less. The limitations of such systems are obvious since they are straightforward anti-growth tests and give no information on tumour-host relationships. False positives result if an agent is very effective at low concentrations but also turns out to be extremely toxic to the host, so that *in vivo* a lethal concentration of the drug in the tumour cannot be achieved at tolerated doses. Conversely false negatives will be encountered if the anti-tumour agent requires activation by the host e. g. liver microsomal metabolism. While such tests have a role to play, for instance in the screening of large numbers of materials only available in small amounts or in assisting isolation and purification of natural products, they can never be used as the sole method of testing for anti-cancer activity. To a certain extent these disadvantages can be overcome by using an *in vivo-in vitro* system where the effect of an agent is compared on the target cell and on normal host cells most sensitive to the drug. Large numbers of platinum complexes may be screened against such systems to provide a lead to the most promising compounds to be prepared on a large scale, but the 'mini' and 'econo' L1210 leukemia tests recently introduced by the National Cancer Institute ([8]), also use only small quantities of material and have the advantages of an *in vivo* system.

Tests on tumour bearing animals have been largely responsible for the discovery of most of the twenty or so agents of established use in the treatment of cancer. Chemically or virally induced tumours have only played a minor role, because the long latent period for tumour induction and the variable time at which tumours appear necessitates keeping large stocks of animals and allows only the testing of small numbers of compounds. It is more economical to use animals with transplanted tumours which may be injected with the test drug the day after transplantation. These tumours are derived from spontaneous or induced tumours and are maintained by successive transplantation from generation to generation. Over the years, a few of these transplanted tumours have been favoured as models for chemotherapy, usually because they were found to be sensitive to one or other class of chemical that subsequently proved to be of clinical value. Examples of such tumours are the Sarcoma 180 and the Walker carcinoma, both of which arose 'spontaneously' and the L1210 leukemia which was induced by application of 20-methylcholanthrene to the skin. The latter tumour is of particular importance since it is the basis of the National Cancer Institutes extensive drug testing and evaluation programme. A retrospective study of agents of established clinical use revealed that the L1210 leukemia system would have detected the majority of these agents with a minimum of false positives, that is, agents active in the animal system but ineffective in man. However, most transplanted tumours used in screening are selecting mainly agents that kill cells in cycle and may not select as active, compounds with selective toxicity for resting, non-mitotic cells. Studies on cell kinetics have shown that most transplanted tumours used in large scale screening have a very high proportion of cells in cycle. This provides an explanation as to why only some human tumours, those with a large proportion of cells in cycle, respond to present day agents and why the limiting toxicity of these agents is damage to normal rapidly growing tissues of the host, such as intestinal mucosa and bone marrow.

Since the most common human cancers contain a large proportion of resting cells which may, under the appropriate conditions, begin to divide, it is obvious that animal models with similar characteristics should be used in screening tests in addition to the present day transplanted tumours.

There has therefore been a re-examination of induced and spontaneous tumours to assess their usefulness as screening systems and the spontaneous AKR leukemia is at present undergoing evaluation. Slower growing transplanted tumours are also under investigation and more recently there have been plans to use human tumours growing in suitably conditioned rodents. It has been possible for some years to grow many types of human tumour in animals, partially deprived of their immune response mechanisms, and *cis*-Pt(II) has shown activity in some preliminary tests against human carcinoma of the colon ([4]) growing in the mouse.

Tumour tests ideally give a quantitative measure of anti-tumour selectivity so that a comparison can be made between members of a chemical series. For solid tumours, effects may be assessed by continual measurement of tumour growth using calipers or by dissection of the tumours some time after treatment, and comparing the tumour weights of treated and control groups. In any test an indication is needed both of the dose to cause a significant anti-tumour effect and the dose that causes host toxicity. To achieve this a series of dose levels are used ranging from lethal to non-tumour inhibitory and selectivity is expressed as a therapeutic index, the ratio of a lethal dose and a tumour effective dose.

For ascites tumours it is usually easier to obtain measurements of tumour cell kill, since most of them show a good inverse relationship between survival time and number of tumour cells injected. The selectivity of a compound is usually expressed by the extension of survival time at the optimum dose and the number of dose levels that gives significant anti-tumour activity.

Table 2 summarises some of the results that have been obtained with *cis*-Pt(II) against a range of animal tumours compared with methotrexate, a clinically used anti-folate and cyclophosphamide, the most extensively used alkylating agent. *Cis*-Pt(II) has a broad spectrum of action and can, in some cases, cause complete tumour regression and cure of the animals. Its effects on the Lewis lung tumour are greater

Table 2. Effects of *cis*-Pt(II) on some experimental tumours

Transplanted tumours	Methotrexate	*cis*-Pt(II)	Cyclophosphamide
RI lymphoma	++	○	○
TLX 5 lymphoma	++	○	○
L 1210 leukemia [a]	++	++	+++
PC 6 plasma cell tumour [b]	○	+++	+++
Sarcoma 180 [c, d]	+	+++	++
Ridgeway sarcoma [e]		+	++
Dunning ascites [f, g]		++	++
Lewis lung tumour [h]	○	++	++
B 16 melanoma	○	++	++
Walker 256	+	++	+++
Walker 256 (resistant to melphalan)	+	○	○
Reticulum cell sarcoma [i]		+	+
Induced tumours			
Lymphocytic leukemia [j]	++	+++	
Myelocytic leukemia [j]	++	+++	
DMBA mammary tumour [k]		+	+

[a] Woodman *et al.* [24].
[b] Connors *et al.* [6].
[c] Clarke [1].
[d] van Camp *et al.* [21].
[e] L. M. van Putten, personal communication.
[f] Kociba *et al.* [14].
[g] Schmidt *et al.* [18].
[h] K. Hellmann, personal communication.
[i] Talley [20].
[j] Leonard *et al.* [15].
[k] Welsch [23].

+++ Complete tumour regression or extension of survival time greater than 200%.
++ Good but incomplete tumour regression or extension of survival time between 50 and 100%.
+ Slight but significant anti-tumour effect.
○ Ineffective.

on the lung metastases than on the primary tumour, in contrast to cyclophosphamide which acts predominantly on the transplanted tumour in the flank. The platinum complex is clearly different from methotrexate in its mechanism of action since lymphomas such as the RI and the TLX5, which are quite sensitive to the anti-folate, do not respond to *cis*-Pt(II). Conversely, the Walker carcinoma and the PC6 plasma cell tumours regress completely after tolerated doses of *cis*-Pt(II) but are hardly affected even by maximal doses of methotrexate. There is a much greater similarity between *cis*-Pt(II) and alkylating agents such as cyclophosphamide. In general, although there are quantitative differences, a tumour sensitive to *cis*-Pt(II) will also respond to cyclophosphamide. Furthermore a tumour (Walker, R., Table 2) with acquired resistance to melphalan and cross-resistant to other alkylating agents is also cross-resistant to *cis*-Pt(II). The platinum complex has other properties reminiscent of alkylating agents. For example, it has a selective effect on DNA synthesis ([11, 12]), cross links DNA ([16]) and is broadly similar to the alkylating agents in its toxicity ([13]) with the exception that *cis*-Pt(II) also causes kidney damage [15]. The toxicity of *cis*-Pt(II) can be reduced by pretreatment with nucleophiles ([5]) although this is no advantage since the anti-tumour action is correspondingly reduced.

Combination Chemotherapy

Most preclinical combination studies have been carried out with the L1210 leukemia which has been shown to be fairly predictable both in selecting clinically useful anti-cancer agents and in investigating the optimal dose schedules to be used in man [10, 7]. Based on this evidence it is reasonable to assume that the L1210 might be a good model for selecting synergistic drug combinations, an assumption which receives support from the finding that effective drug combinations used in the treatment of acute lymphocytic leukemia show therapeutic synergism in the L1210 leukemia system [9].

Combinations of *cis*-Pt(II) with each of a number of different agents have been tested against the early L1210 leukemia and the most effective of these further evaluated against the advanced tumour which is less sensitive to chemotherapy [24]. These studies will no doubt be extended to other tumour systems and there is already some preliminary data on the effect of two drug combinations containing *cis*-Pt(II) on the sarcoma 180 [21].

A series of two drug combinations tested by Woodman *et al.* showed that some, but not all, combinations were therapeutically synergistic. Some of these results are summarised in Table 3. Cyclophosphamide, and its analogue isophosphamide, were very effective in inhibiting the growth of the L1210 leukemia but were not curative since all animals eventually succumbed to the disease before sixty days. There was some additive toxicity when they were used in combination with *cis*-Pt(II) since the dose levels of each agent used in the combination had to be reduced. However, the anti-tumour effect measured in terms of survival time or in numbers of tumour-free animals at 60 days was better in the combination than with either drug alone. Similar synergism is seen with ICRF 159 or emetine. These last two agents differ fundamentally from *cis*-Pt(II) in their mechanism of action so that there is no addition of toxicity, the optimal single dose of each compound being used in the combination. In both cases there was an increase in life span and in addition the ICRF, *cis*-Pt(II) combination gave 25% long-term survivors. In the case of BCNU and *cis*-Pt(II), however, there was no evidence of therapeutic synergism, the single optimal dose of BCNU being apparently more effective than the combination with *cis*-Pt(II). Extension of these tests to the advanced L1210 leukemia with ICRF 159 and *cis*-Pt(II) showed that the synergism of the combination was retained against the systemic disease.

Table 3. Effects of *cis*-Pt(II) alone and in combination with various anti-cancer agents. From the data of WOODMAN *et al.* [24]

Compounds	Treatment schedule	Optimal dose (mg/kg)	% ILS	60 day survivors
cis-Pt(II)	Single day 3	14	55	0/8
Isophosphamide	Single day 3	300	100	0/8
Isophosphamide + *cis*-Pt(II)	Single day 3	180+3	469	4/8
cis-Pt(II)	Single day 3	12	27	0/8
Cyclophosphamide	Single day 3	200	140	0/8
Cyclophosphamide + *cis*-Pt(II)	Single day 3	200+6	495	4/8
cis-Pt(II)	days 1, 5, 9	4	78	0/8
ICRF 159	days 1, 5, 9	200	122	0/8
ICRF 159 + *cis*-Pt(II)	days 1, 5, 9	200+4	300	2/8
cis-Pt(II)	days 1, 5, 9	4	55	0/8
Emetine	days 1, 5, 9	15	61	0/8
Emetine + *cis*-Pt(II)	days 1, 5, 9	15+4	122	0/8
cis-Pt(II)	Single day 5	8.4	55	0/8
BCNU	Single day 5	40	560	6/8
BCNU + *cis*-Pt(II)	Single day 5	20+3.9	395	4/8

Essentially similar results have been obtained by SPEER *et al.* [19] using the same system. A number of therapeutically synergistic combinations were demonstrated, in particular, *cis*-Pt(II) with either cyclophosphamide, 5-fluorouracil or thioguanine. Combinations of *cis*-Pt(II) with methotrexate, vinca alkaloids and hydroxyurea, were also synergistic but less selective. The increased effectiveness of cyclophosphamide with *cis*-Pt(II) has also been observed on the sarcoma 180 ([21]) and indicates that, despite the rather similar biological properties of the two compounds, they may still be used with advantage in combination. These findings are supported by the preliminary clinical observations of Wiltshaw (this journal) that ovarian carcinoma resistant to alkylating agents would still respond to *cis*-Pt(II).

Analogues of *cis*-Pt(II)

The narrow structure activity relationships of the platinum complexes became clear from the testing of the very first analogues of *cis*-Pt(II). It was soon established that the *trans* analogue was inactive and that substitution of platinum by other heavy metals did not give active compounds (Table 4). More recently some activity has been observed with some rhodium (Table 4) and palladium complexes (P. CONRAN, this journal). The most effective derivatives appear to be *cis*-dichloro platinum compounds containing substituted amines, although the nature of the amine is critical. More detailed analyses of the structure activity relationships are given elsewhere ([2, 3, 6]) but a few examples will be dealt with here.

The good anti-tumour properties of *cis*-Pt(II) against a plasma cell tumour are lost when a methyl group is placed on the nitrogen atoms, even though there is no reduction of toxicity (Table 5). A β-chloroethyl substituent on the nitrogen does lower

Table 4. Effect of some metal complexes on mice bearing the PC6 plasma cell tumour

	LD_{50}	ID_{90}	TI
$(H_3N)(H_3N)Pt(Cl)(Cl)$	13.0	1.6	8.1
$(Cl)(H_3N)Pt(NH_3)(Cl)$	27.0	> 27.0	<1.0
$(H_3N)(H_3N)Pd(Cl)(Cl)$	50.0	> 50.0	<1.0
$(H_3N)(H_3N)(H_3N)Ir(Cl)(Cl)(Cl)$	1500	>1500	<1.0
$(H_3N)(H_3N)(H_3N)Ru(Cl)(Cl)(Cl)$	132	> 132	<1.0
$(H_3N)(H_3N)(H_3N)Rh(Cl)(Cl)(Cl)$	225	86	2.6

The ID_{90} is the minimum dose that causes complete tumour regression. The ratio of the LD_{50} and ID_{90} is the therapeutic index (TI).

Table 5. Effect of some analogues of *cis*-Pt(II) on the plasma cell tumour

	LD_{50}	ID_{90}	TI
$(H_3N)(H_3N)Pt(Cl)(Cl)$	13.0	1.6	8.1
$(CH_3NH_2)(CH_3NH_2)Pt(Cl)(Cl)$	18.5	18.5	1.0
$(Cl \cdot CH_2CH_2NH_2)(Cl \cdot CH_2CH_2NH_2)Pt(Cl)(Cl)$	45.0	17.5	2.6
$(CH_3 \cdot CN)(CH_3 \cdot CN)Pt(Cl)(Cl)$	27.0	27.0	1.0
$(C_6H_5CN)(C_6H_5CN)Pt(Cl)(Cl)$	90.0	90.0	1.0

The ID_{90} is the minimum dose that causes complete tumour regression. The ratio of the LD_{50} and ID_{90} is the therapeutic index (TI).

Table 6. Effect of some analogues of *cis*-Pt(II) on the PC6 plasma cell tumour

	LD_{50} (mg/kg)	ID_{90} (mg/kg)	TI
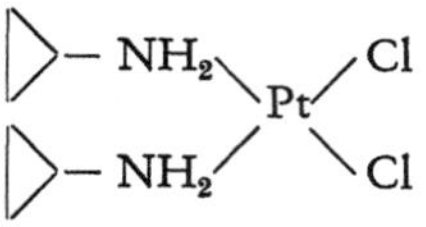	56.6	2.3	24.6
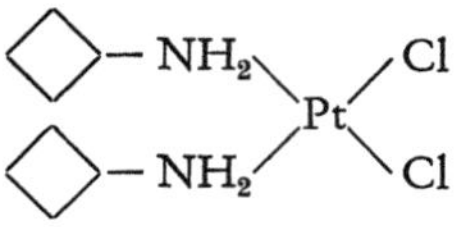	90.0	2.9	31.0
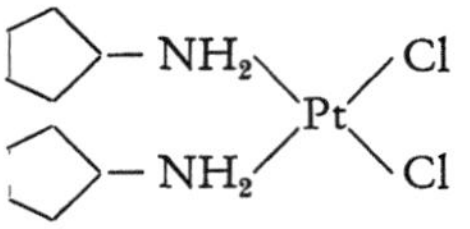	480	2.4	200
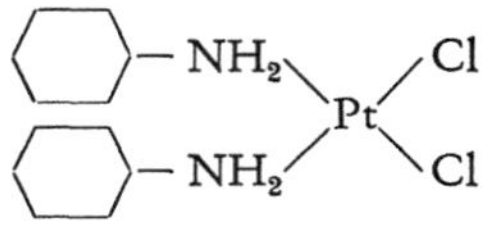	3200	12.0	267
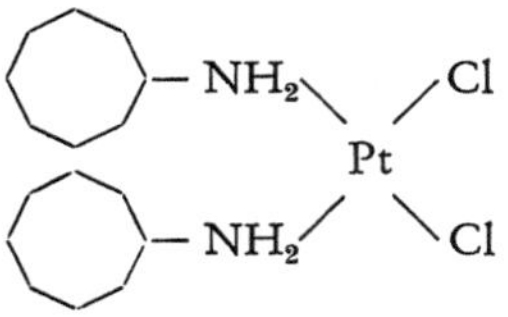	135	59	2.3

The ID_{90} is the minimum dose that causes complete tumour regression. The ratio of the LD_{50} and ID_{90} is the therapeutic index (TI).

the toxicity but decreases selectivity, while more fundamental changes in the nature of the nitrogen donor as in the aceto and benzonitriles (Table 5) cause complete loss of anti-tumour action. At least two series of platinum complexes contain members which are significantly more active against the plasma cell tumour than *cis*-Pt(II). In one series where the ammonia ligands are replaced by compounds containing cyclic nitrogen atoms, two agents, *cis*-dichlorobis(ethyleneimine)platinum(II) and *cis*-dichlorobis(pyrollidine)platinum(II) are two to three times more selective than the parent compound. Even in this series, however, small changes in structure, e. g. substitution of the ethyleneimine, or replacement of pyrollidine by morpholine, can lead to complete loss of activity.

The second series of platinum complexes which has members significantly better than the parent compound are *cis*-dichloro derivatives co-ordinated with an aromatic or alicyclic diamine or two alicyclic amines. The alicyclic series is shown in Table 6. As the molecular weight of the alicyclic substituents increases, so there is an increase in therapeutic index. The anti-tumour potency remains at about 2.5 mg/kg until the cyclobutyl compound, but thereafter it decreases, although there is still an increase in the therapeutic index because the LD_{50} increases to a greater extent. An optimum molecular size of the alicyclic substituents exists since the cyclooctyl member is only a weak anti-tumour agent because of both high toxicity and poor anti-tumour potency.

In conclusion, the platinum complexes are still of considerable interest because of their wide spectrum of anti-tumour action in laboratory animals. Although generally similar to the nitrogen mustards, they differ in some respects and the structure activity relationships of the series implies that new derivatives may yet be synthesised with even higher anti-tumour selectivity. The kidney toxicity of the first members of the series and the insolubility of some of the most active compounds are still two problems that remain to be overcome but the discovery of the highly water-soluble complexes of platinum dichlorodiamines with pyrimidines is just one way in which this may be achieved.

References

1. CLARKE, D. A.: Mouse Sarcoma 180. Cancer Res. Suppl. **3**, 15 (1955).
2. CLEARE, M. J., HOESCHELE, J. D.: Anti-tumour platinum compounds. Relationship between structure and activity. Platinum Metals Review **17**, 2 (1973).
3. CLEARE, M. J., HOESCHELE, J. D.: Studies on the anti-tumour activity of Group VIII transition metal complexes. Part I. Platinum(II)complexes. Bio inorganic Chemistry **2**, 187 (1973).
4. COBB, L. M., MITCHLEY, B. C. V.: Personal communication.
5. CONNORS, T. A.: Chemotherapy of animal tumours with platinum compounds. In: Advances in anti-microbial and anti-neoplastic chemotherapy, Vol. 2, p. 237. München, Berlin, Wien: Urban & Schwarzenberg 1972.
6. CONNORS, T. A., JONES, M., ROSS, W. C. J., BRADDOCK, P. D., KHOKHAR, A. R., TOBE, M. L.: New platinum complexes with anti-tumour activity. Chem.-Biol. Interactions **5**, 415 (1972).
7. GOLDIN, A.: Rationale of combination chemotherapy based on preclinical experiments. Cancer Chemother. Rep. (part 3) **4**, 189 (1973).
8. GOLDIN, A., MANTEL, N.: Cancer Chemother. Rep. (in press).
9. GOLDIN, A., SANDBERG, J. S., HENDERSON, E. S., NEWMAN, J. W.: The chemotherapy of human and animal acute leukaemia. Cancer Chemother. Rep. **55**, 309 (1971).
10. GOLDIN, A., SERPICK, A. A., MANTEL, N.: Experimental screening procedures and clinical predictability value. Cancer Chemother. Rep. **50**, 173 (1966).
11. HARDER, H. C., ROSENBERG, B.: Inhibitory effects of anti-tumour platinum compounds on DNA, RNA and protein synthesis in mammalian cells *in vitro*. Int. J. Cancer **6**, 207 (1970).
12. HOWLE, J. A., GALE, G. R.: *Cis*-dichlorodiammine platinum(II) persistent and selective inhibition of DNA synthesis *in vivo*. Biochem. Pharmacol. **19**, 2757 (1970).
13. KOCIBA, R. J., SLEIGHT, S. D.: Acute toxicologic and pathologic effects of *cis*-diammine dichloroplatinum (NSC–119 875) in the male rat. Cancer Chemother. Rep. (part 1) **55**, 1 (1971).
14. KOCIBA, R. J., SLEIGHT, S. D., ROSENBERG, B.: Inhibition of dunning ascitic leukaemia and Walker 256 carcinoma with *cis*-diammine-dichloroplatinum (NSC-119 875). Cancer Chemother. Rep. **54**, 325 (1970).
15. LEONARD, B. J., ECCLESTON, E., JONES, D., TODD, P., WALPOLE, A.: Anti-leukaemic and nephrotoxic properties of platinum compounds. Nature (Lond.) **234**, 43 (1971).
16. ROBERTS, J. J., PASCOE, J. M.: Cross-linking of complementary strands of DNA in mammalian cells by anti-tumour platinum compounds. Nature (Lond.) **235**, 282 (1972).
17. ROSENBERG, B., VAN CAMP, L., TROSKO, J. E., MANSOUR, V. H.: Platinum compounds: a new class of potent anti-tumour agent. Nature (Lond.) **222**, 385 (1969).

18. SCHMIDT, L. H., FRADKIN, R., SULLIVAN, R., FLOWERS, A.: Comparative pharmacology of alkylating agents. Cancer Chemother. Rep. Suppl. **2**, pt. I-III (1965).
19. SPEER, R. J., LAPIS, S., RIDGWAY, H., MEYERS, T. D., HILL, J. M.: *Cis* platinous diammino dichloride (PDD) in combination therapy of leukaemia L 1210. Advances in anti-microbial and anti-neoplastic chemotherapy. Vol. 2, p. 253. München, Berlin, Wien: Urban & Schwarzenberg 1972.
20. TALLEY, R. W.: Chemotherapy of a mouse reticulum cell sarcoma. Proc. Amer. Ass. Cancer Res. **11**, 78 (1970).
21. VAN CAMP, L., ROSENBERG, B.: Combination therapy of *cis*-dichloro-diammineplatinum(II) with cytoxan against the sarcoma 180 tumor in Swiss white mice. In: Advances in anti-microbial and anti-neoplastic chemotherapy. Vol. 2, p. 239. München, Berlin, Wien: Urban & Schwarzenberg 1972.
22. VENDITTI, J. M., GOLDIN, A.: Drug synergism in anti-neoplastic chemotherapy. In: Ed. GOLDIN and HAWKING. Advances in Chemotherapy. Vol. 1, p. 397. New York: Academic Press 1964.
23. WELSCH, C. W.: Growth inhibition of rat mammary carcinoma induced by *cis* platinum diamminodichloride(II). J. nat. Cancer. Inst. **47**, 1071 (1971).
24. WOODMAN, R. J., SIRICA, A. E., GANG, M., KLINE, I., VENDITTI, J. M.: The enhanced therapeutic effect of *cis*-platinum(II) diamminodichloride against L 1210 leukaemia when combined with cyclophosphamide or 1,2-bis(3,5-dioxo-piperazine-1-yl) propane or several other anti-tumour agents. Chemotherapy **18**, 169 (1973).

Pharmacokinetics of Platinum Compounds

P. B. CONRAN (Rapporteur)

1. Metabolism of platinum(^{14}C)ethylenediamine dichloride in the rat.
TAYLOR, D. M., JONES, J. D., ROBINS, A. B., Institute of Cancer Research, Sutton, Surrey/England.
2. Radiopharmacokinetics of $^{195m}Pt^{m}$ *cis*-diamminedichloroplatinum(II).
WOLF, W., MANAKA, R. C., University of Southern California, Los Angeles, California, USA.
3. The cytotoxic effect of *cis*-dichlorodiammineplatinum(II) *in vitro* and its potency against experimental animal tumours.
SVEC, F., KOVARIK, J., Cancer Research Institute Cs. Armady. Bratislava, Czechoslovakia.
4. Note on the effect of *cis*-dichlorodiammineplatinum(II) on the intestinal epithelium of the rat.
RYPACEK, F., CERKASOV, J., DROBNIK, J., Institute of Macromolecular Chemistry. Prague, Czechoslovakia.
5. *Cis*-dichlorodiammineplatinum(II) (NSC 119 875). Preclinical toxicological evaluation of i.v. injection in dogs, monkeys and mice. U.S. Government Res. Report MR1-CC02-71-19-1971 and MRI-CC05-71-45.
SCHAEPPI, U., HEYMAN, I. A., FLEISCHMAN, R. W., ROSENKRANZ, H., LLIEVSKI, V., PHELAN, R., COONEY, D. A., DAVIS, R. D., Mason Research Institute. Worcester, Mass./USA.
6. Urinary lactic dehydrogenase (ULDH) and alkaline phosphatase (UAP) as indicators of renal toxicity in rhesus monkeys treated with *cis*-dichlorodiammineplatinum(II).
HEYMAN, I., FLEISCHMAN, R. W., SCHAEPPI, U., COONEY, D., DAVIS, R., National Cancer Institute, NIM. Bethesda, Md./USA.
7. *Cis*-dichlorodiammineplatinum(II) induced ototoxicity in guinea pigs.
STADNICKI, S. W., FLEISCHMAN, R. W., SCHAEPPI, U., ETHIER, M., Mason Research Institute, Worcester, Mass./USA.
8. Immunosuppression with platinum compounds.
KHAN, A., Wadley Institutes of Molecular Medicine. Dallas, Texas/USA.
9. *Cis*-dichlorodiammineplatinum(II): suppression of adjuvant induced arthritis in rats.
BOWEN, J. R., GALE, G. R., GARDNER, W. A., JR., BONNER, W. M., U.S.V.A. Hospital, Charleston, S.C./USA.
10. Serum electrophoretic patterns in Swiss mice bearing S 180 and treated with *cis*-Pt(II).

Pant, H. C., Van Camp, L., Peresie, H. J., Sodhi, A., Rosenberg, B., Department of Biophysics, Michigan State University, East Lansing/USA.

11. The enhanced therapeutic effect of *cis*-platinum(II) diamminedichloride against L1210 leukemia when combined with cyclophosphamide or 1,2-bis(3,5-dioxopiperazine-1-yl)propane or several other anti-tumour agents.
Woodman, R. J., Sirica, A. E., Gang, M., Kline, I., Venditti, J. M., National Cancer Institute, Bethesda, Maryland 20014/USA.
12. Platinum compounds and their anti-tumour effects in mice.
Ridgway, H., Lapis, S., Hall, L. M., Stewart, D. P., Hill, J. M., Speer, R. J., Wadley Institutes of Molecular Medicine. Dallas, Texas/USA.
13. Enhanced *in vivo* activity of combined *cis*-Pt(II)-diammine-dichloride (NSC 119 875) and gamma-radiation therapy.
Wodinsky, I., Swiniarsky, J., Kensler, C. J., Venditti, J. M., Arthur D. Little, Inc., Cambridge, Massachusetts/USA.
14. The role of host defenses in mediating tumor regressions in mice treated with *cis*-Pt(II).
Conran, P., University of Connecticut Health Center. Farmington, Connecticut/USA.
15. Chemotherapy of platinum amine complexes against ascites S-180 tumors in mice.
Peresie, H. J., Fischer, R. G., Jr., Van Camp, L., Rosenberg, B., Department of Biophysics, Michigan State University. East Lansing/USA.

The recent interest in the use of platinum compounds as antitumor agents precipitated a need for information on the pharmacokinetics and tissue distribution of these drugs. The most effective platinum compound showing antitumor activity is *cis*-diamminodichloride platinum(II) [*cis*-Pt(II)]. Unfortunately, however, it can only be labeled with platinum radionuclides of short half-life e. g. ^{195m}Pt. Another less effective compound, platinum ethylenediamine dichloride [Pt(en)Cl_2] can be labeled with either radioactive carbon (^{14}C) on the ethylenediamine moiety (en) or with both ^{197m}Pt and ^{14}C-en. The problem with the latter compound, however, is the lack of information regarding whether or not the en moiety remains with the parent compound and can be used as a monitor of drug distribution. The other remaining question is whether the distribution and kinetics of Pt(en)Cl_2 is similar to that of *cis*-Pt(II).

Addressing themselves to these questions, Taylor *et al.* studied the tissue and subcellular distribution of ^{195m}Pt(^{14}C-en)Cl_2 and compared it to that of *cis*-^{195m}Pt(II). The animals chosen for this study were rats, some of which were implanted with either the BICR/A3 osteogenic sarcoma or the BICR/M2 fibrosarcoma.

The results of these experiments indicated that the tissue distribution and concentration of ^{195m}Pt and ^{14}C-en were virtually identical. This suggested that Pt(en)Cl_2 did not dissociate to any marked extent *in vivo* and that the distribution of ^{14}C was a good measure of the distribution of the parent compound.

The maximum concentration of the drug in the tumors was approximately 0.2% of the injected dose. This suggested a lack of specific uptake by the tumor. It was also of interest to note that the concentration of drug varied from one area of the tumor to another by factors of up to 3 : 1.

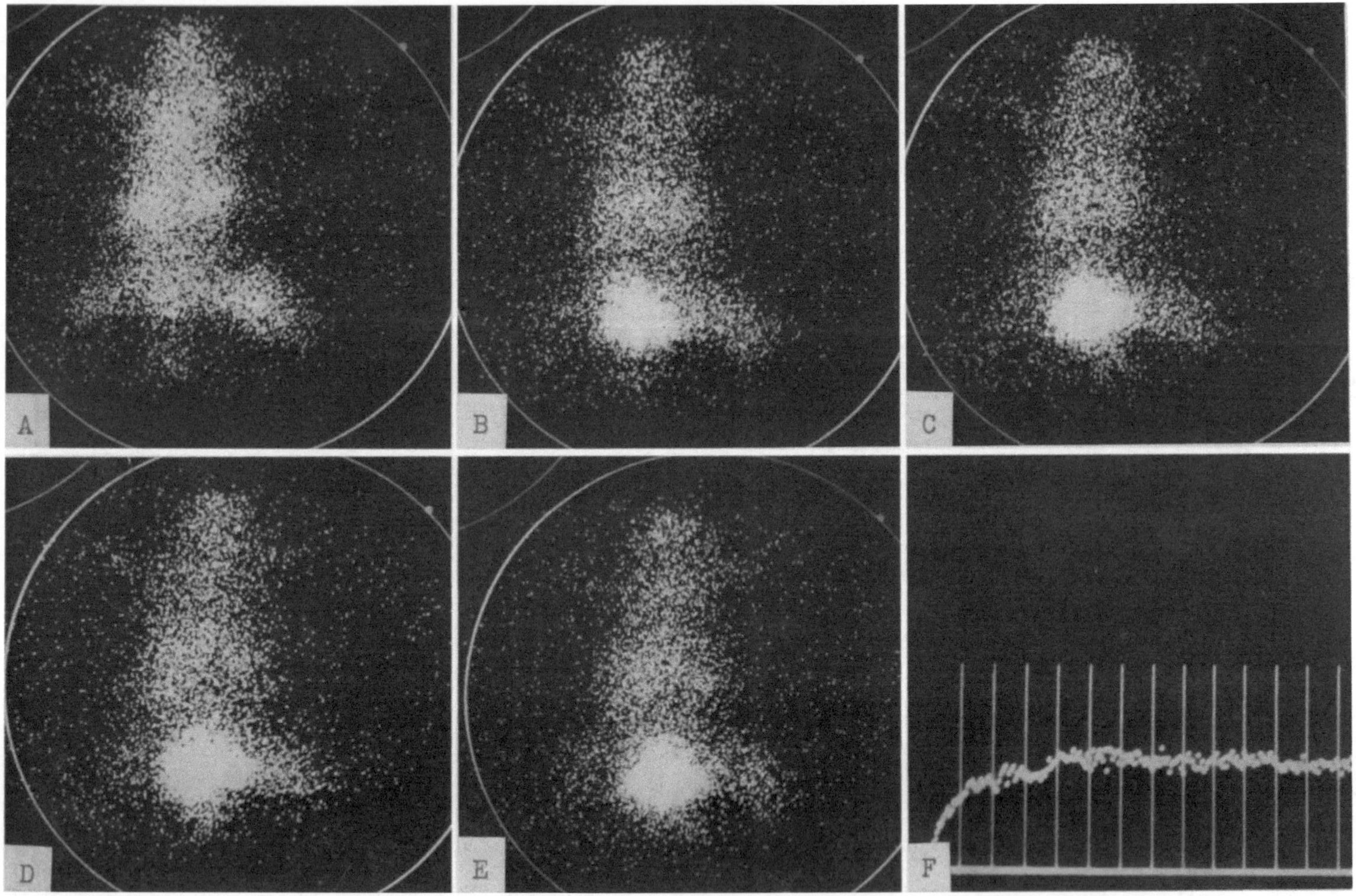

Fig. 1 a to f. γ camera pictures of rats bearing the Walker 256 tumor, following injection of $^{195}Pt^{m}$-*cis*-diamminedichloroplatinum(II). (A) t = 0; (B) t = 0.5 h; (C) t = 1.0 h; (D) t = 2.0 h; (E) t = 3.0 h; (F) activity v. s. time, bladder

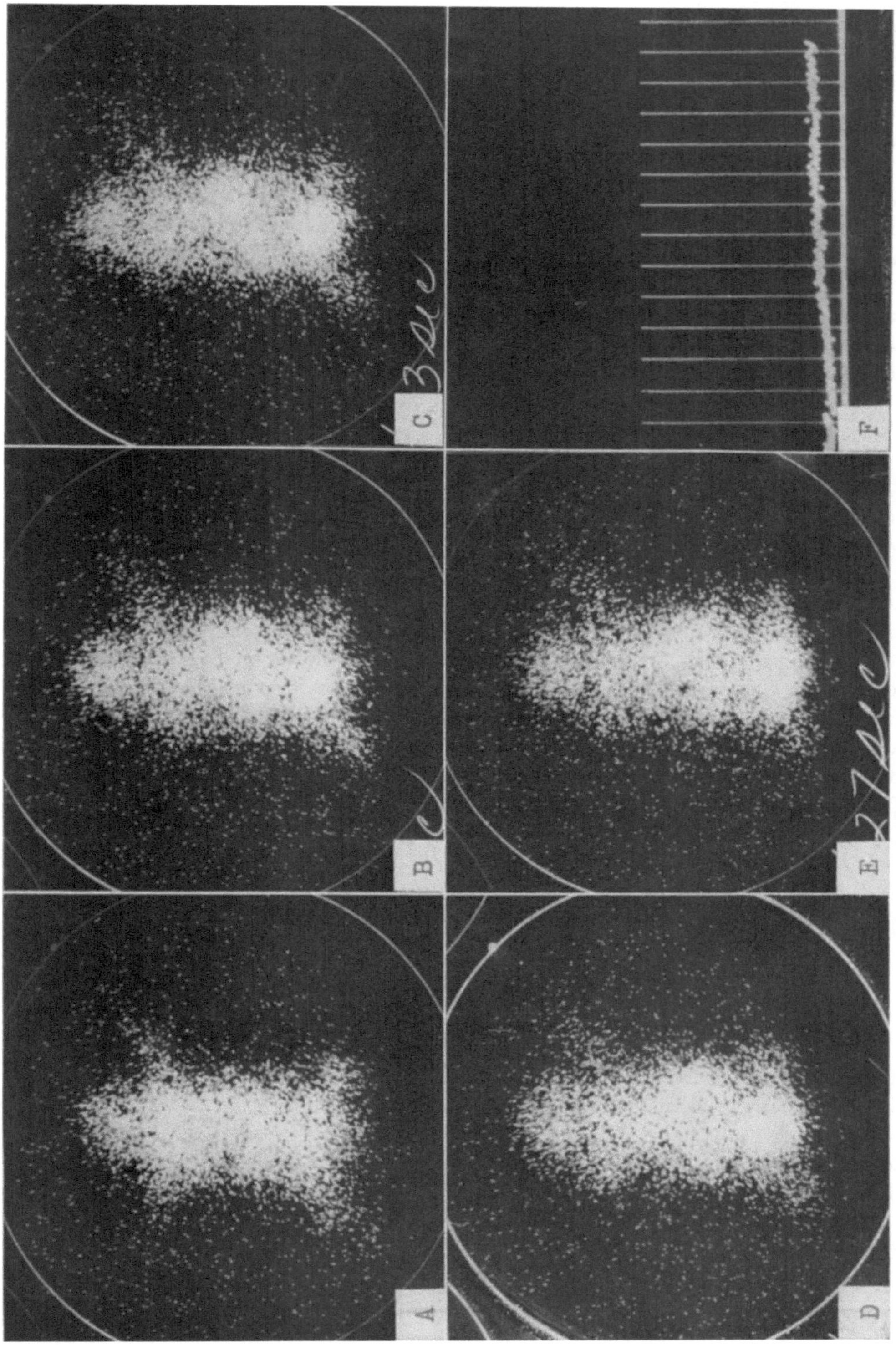

Fig. 2 a to f. γ camera pictures of control rats, following injection of $^{195}Pt^{m}$-*cis*-diamminedichloroplatinum(II). (A) t = 0; (B) t = 0.5 h; (C) t = 1.0 h; (D) t = 2.0 h; (E) t = 3.0 h; (F) activity v. s. time, bladder

The distribution and retention of radiolabeled *cis*-Pt(II) was similar to that of radiolabeled Pt(en)Cl_2. Approximately 55% of both drugs was excreted via the kidneys by 24 h post-injection. The uptake of *cis*-Pt(II) in the kidneys was four times greater and in the liver two times greater than that of Pt(en)Cl_2. On the other hand, the plasma half-time retention of *cis*-Pt(II) was shorter (1.9 days) than that of Pt(en)Cl_2 (2.5 days).

WOLF and MANAKA studied the kinetics of *cis*-^{195m}Pt(II) in rats bearing the Walker 256 carcinosarcoma and compared it to that in non-tumor bearing controls. Like TAYLOR *et al.* these authors found a significant uptake of the drug in the kidney and liver but no selective uptake by the tumor.

The concentration of the drug in the kidneys of animals bearing tumors was more variable than that of the controls. The authors postulated that this might have been due to an effect by the tumor on renal function. Thus, the variability may be a reflexion of the tumor status at the time of monitoring.

It was of interest to note that commencing at 48 h and continuing through 72 h after administration, there was a marked increase in the concentration of the drug in virtually all tissues evaluated. This apparent delayed release suggested the presence of a "drug reservoir". The authors proposed that there are three potential sites for drug accumulation to occur: 1. the gastrointestinal tract, 2. the liver, or 3. fluids associated with a generalized edematous process.

When the excretion and localization of *cis*-^{195m}Pt(II) were compared 3 h after administration between the two animal groups, it was found that the tumor-bearing animals cleared the first fraction of the drug approximately ten times faster than the control animals (Figs. 1, 2). The excretion rates were slower in both groups after four hours. The marked difference in excretion rate of the first fraction between the two groups was exemplified in one rat in which it was assumed that the implanted tumor had not grown. The kinetics of drug excretion in this animal, however, suggested the presence of a tumor. Upon dissection a small, but definite tumor was found. This observation could prove to be of diagnostic importance.

Toxicology of Platinum Compounds

In vitro Studies

Numerous *in vitro* studies have been performed to ascertain the effects of platinum compounds on DNA, RNA and protein synthesis. On the other hand, there is little information regarding their affect on other metabolic processes.

SVEC and KOVARIK studied the affects of varying concentrations of *cis*-Pt(II) on cellular division and oxygen (O_2) consumption in a line of Hela cells. The results of these experiments are summarized in Table 1.

In those cells in which the O_2 consumption was decreased, vital staining did not disclose significant differences in viability between treated and non-treated cells. Thus, the authors suggested that *cis*-Pt(II) induced respiratory inhibition could be a consequence of the compound's capability to interfere with cellular protheosynthesis. Under these experimental conditions, there was a parallelism between inhibition of

Table 1. The effect of *cis*-Pt(II) on Hela cell viability, proliferation and oxygen consumption

Concentration *cis*-Pt(II)	Cellular division	Oxygen consumption
Controls	100%	8.72 µl $O_2/1\times10^6$ cells
5×10^{-4} m	0% viability, 3—5 h	
1×10^{-4} m	2.7% viability 100% inhibition	
1×10^{-5} m	71% inhibition	3.29 µl $O_2/1\times10^6$ cells
1×10^{-6} m	45% inhibition	

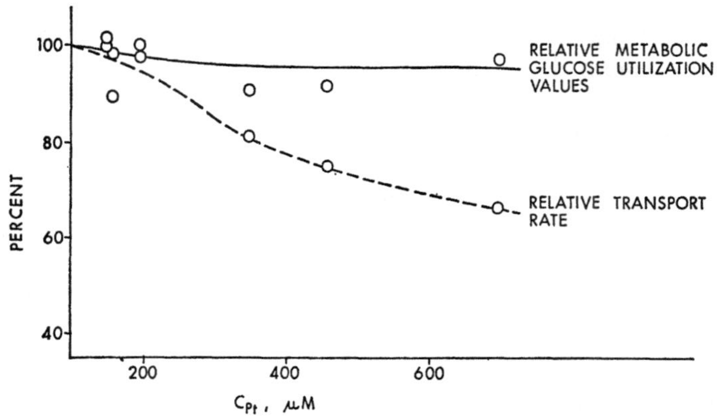

Fig. 3. The effects of *cis*-Pt(II) on glucose metabolism and transport in rat intestinal epithelium

mitosis and the inhibition of metabolic pathways. Thus, the authors concluded that the measurement of both of these parameters could serve as a suitable supplement to *in vivo* tests for the prediction of the drug's cytostatic activity.

The authors reported that the *cis*-Pt(II) inhibitory dose 50 (ID_{50}) for Hela cells was 1.5×10^{-6} M which is in the range of that found with alkylating agents. This dose is approximately 1000 times lower than that found with antimetabolites.

The effect of *cis*-Pt(II) on glucose metabolism and transport in rat intestinal epithelium was studied by Rypacek *et al.* Segments of small intestine, 3—4 cm in length, cut at a distance of 30 cm from the pylorus were used in the study. The segments were filled with glucose or glucose and *cis*-Pt(II) and incubated in test tubes containing similar media.

The rate of active transport of glucose was expressed as an increase in glucose content in the serous volume within the intestinal segment. The metabolic utilization of glucose was determined by measuring the total loss of glucose in the serous and mucous volume.

The results of this experiment are depicted in Fig. 3. While *cis*-Pt(II) did not affect glucose metabolism, it did decrease the transport of glucose. This decrease was dose dependent. A 30% reduction was attained at a dose level of 700 µm, a dose which is 17 times that of the lethal dose (LD_{50}) in rats. Consequently, the authors concluded that glucose transport is moderately affected by *cis*-Pt(II) at high doses.

In vivo Studies

Schaeppi *et al.* studied the toxicological effects of intravenous (IV) injections of *cis*-Pt(II) in mice, dogs and monkeys. The LD_{50} for a single dose IV injection in Swiss mice was 13.4 mg per kg for females. The minimum lethal dose (MLD) for the dog was a single IV injection of 2.5 mg per kg or five consecutive daily injections of 0.75 mg per kg. The MLD for monkeys was five consecutive daily doses of 2.5 mg per kg.

Treatment utilizing the MLD in dogs and monkeys elicited severe morbidity within five to seventeen days. The speed of onset was related to the dose administered. Toxicity included severe hemorrhagic enterocolitis, severe hypocellularity of the bone marrow and lymphoid tissues, marked renal tubular necrosis in dogs and nephrosis in monkeys.

The renal lesions were the most severe toxic changes in survivors. Clinically, renal toxicity was characterized by azotemia, hypochloremia, proteinuria and the appearance of erythrocytes and leukocytes in the urine. There was an increased 24 hour excretion of urinary lactic dehydrogenase (ULDH) in dogs and a protracted polyuria in monkeys.

Occasionally, hypocalcemia, myocarditis (monkeys), pancreatis (dogs) and testicular degeneration was observed.

The results of this study confirmed previous reports that treatment with *cis*-Pt(II) primarily affects tissues with a high rate of cell division e. g. intestines, bone marrow, lymphoid tissue and testes and tissues which are known for primary excretion of the compound e. g. intestines and kidneys.

Schaeppi *et al.* measured ULDH levels as a clinical monitor of renal toxicity. The value of ULDH and urinary alkaline phosphatase (UAP) excretion rates as a measure of renal pathology was evaluated by Heyman *et al.* The UAP levels remained unchanged in Rhesus monkeys injected with *cis*-Pt(II) for five days at a dose of 0.625 mg/kg day. The excretion of ULDH, however, increased three to ten-fold commencing three to eight days after treatment was initiated; reaching its highest level between seven and 21 days. At the termination of the experiment (54 days), there was a reversal of ULDH activity which correlated with the absence of renal lesions.

Thus it was concluded that ULDH excretion can be used as a monitor of *cis*-Pt(II) induced renal toxicity.

Aside from renal toxicity, one of the major problems associated with the use of *cis*-Pt(II) in humans has been ototoxicity. In an attempt to illucidate the mechanism by which this occurs, Stadnicki *et al.* studied the effects of various doses of the drug in the guinea pig. Ototoxicity was evaluated by measuring the disappearance of Preyer's reflex in response to pure tones of 5, 7 and 10 kHz.

The results of this experiment are summarized in Table 2. Phase contrast microscopy of surface preparations from the organ of Corti revealed a loss of outer hair

Table 2. The effects of *cis*-Pt(II) on auditory function in guinea pigs

Dose	Deafness [b]	Mortality [b]
9 mg/kg	4/10	2/10
12 mg/kg	4/10	7/10
1.5 mg/kg [a]	13/13	4/13

[a] 4—16 daily doses administered 5 X's/week.
[b] Numerator=number of animals with characteristic; denominator=total number of animals.

cells primarily in the basal turn of the cochlea. Repeated treatment caused disappearance of entire rows of outer hair cells with occasional survivors. A few outer pillar cells were lost as well. Single dose treatment produced a scattered pattern of outer hair cell loss. The lesions in all instances were characterized by degeneration, complete cytolysis and fibrosis.

None of the doses or dose schedules produced loss of inner hair cells or vestibular toxicity.

Immunology

KHAN *et al.* have previously reported that platinum compounds are immunosuppressive agents. They found that *cis*-diamminotetrachloride platinum(IV) [*cis*-Pt(IV)] inhibits phytohemagglutinin (PHA) induced lymphocyte blastogenesis and inhibits antibody plaque-forming spleen cells in mice. In addition, they reported that *cis*-Pt(II) inhibits PHA induced lymphocyte blastogenesis *in vitro* and *in vivo* (human patients), suppresses graft rejection and graft versus host reactions in mice, and inhibits antibody plaque-forming spleen cells in mice.

Recently, KHAN studied the effects of dichloro-bis-(cyclohexylamine)-platinum(II) (DP) on immune responses. This compound is effective against plasma cell tumors, but not against L 1210 leukemia.

The results of the affect of DP on antibody plaque-forming spleen cells in mice are summarized in Table 3. It is of interest to note that while DP exhibited immunosuppressive capabilities, the sesame seed oil in which the compound was suspended acted as a stimulant to immune responses.

When 500 mg/kg of DP was administered prior to, concomitantly or after sheep red blood cells, it was found that maximum immunosuppression occurred when the compound was given either seven days or three days prior to the antigen. On the other hand, *cis*-Pt(II) produced maximum immunosuppression when given concomitant with or after the antigen.

Preliminary studies indicate that delayed hypersensitivity in rabbits is not suppressed by DP at 500 mg/kg.

The results of these experiments indicate that the mode of action of DP may differ from that of *cis*-Pt(II) and *cis*-Pt(IV). The authors suggested that it may exert its maximum effect on B lymphocytes, sparing T cells.

A possible clinical application of the immunosuppressive capabilities of *cis*-Pt(II) was reported by BOWEN *et al.* They found that doses of 1.0 and 0.75 mg/kg per day,

Table 3. Suppression of PFC by dichloro bis (cyclohexylamine) platinum(II) (DP) in mice

Group	Immunization	Treatment DP (mg/kg)	PFC/$10^6 \pm$ SE	PFC$\times 10^8$/spleen $\pm$ SE
A	—	—	1 ± 0.4	1 ± 0.3
B	SRBC	—	145 ± 25	127 ± 21
C	SRBC	SSO	251 ± 63 [a]	186 ± 26 [a]
D	SRBC	250	26 ± 9 [a]	11 ± 4 [a]
E	SRBC	500	23 ± 14 [a]	11 ± 9 [a]

[a] Significant differences from group B ($P<0.01$).
SSO = sesame seed oil.

given on days 10—20 of the experiment, diminished the inflammatory response due to adjuvant-induced arthritis in rats.

In this study it was noted that arthritic rats treated with *cis*-Pt(II) manifested vascular congestion and focal hemorrhage in the bone marrow. Whether the drug or the combination of the adjuvant-induced arthritis and the drug caused this condition is not known.

PANT *et al.* studied serum protein electrophoretic patterns in Swiss mice bearing solid Sarcoma 180 (S-180) tumors and compared them to those of non-tumor bearing animals. In addition, some of the animals in each group were treated with 7 mg/kg of *cis*-Pt(II).

The results of this experiment indicated that there were changes in the intensity of the α_2 region. These changes were directly related to the status of tumor growth. As the tumor size increased, the intensity of the α_2 region increased. On the other hand, as the size of the tumor decreased, the intensity of the α_2 region decreased. With complete tumor regression, the α_2 region returned to control levels. These changes were similar in both spontaneous and *cis*-Pt(II) induced tumor regressions.

The nature of the α_2 protein associated with tumor growth was not determined.

No changes were observed in the β or γ regions.

Cancer Chemotherapy with Platinum Compounds

Combination of Platinum Compounds with Other Chemotherapeutic Agents

While *cis*-Pt(II) alone has shown antitumor activity against a wide range of animal tumors, many investigators have found that the combination of the platinum compound and various other cancer chemotherapeutic agents results in therapeutic potentiation. Therapeutic potentiation is antitumor activity promoted by the combination of two or more drugs which surpasses that of the most active drug when used alone. A summary of the drugs that have been used successfully in combination with *cis*-Pt(II) is found in Table 4.

WOODMAN *et al.* found that combination of *cis*-Pt(II) and cyclophosphamide, isophosphamide, ICRF-159 or emetine yielded increases in life span that were at least double that achieved with the more effective of the individual drugs against early

Table 4. Combinations of *cis*-Pt(II) with other agents reported to be therapeutically potentiating against leukemia L1210 [a]

cis-Pt(II) with:	
Cyclophosphamide	ICRF-159
Isophosphamide	5-Fluorouracil
BIC	6-Thioguanine
Methotrexate	Vincristine
5-Azacytidine	Vinblastine
Cytosine arabinoside	Camptothecin
Phosphoramide mustard	Emetine

[a] From F. S. Schabel, Jr. (in press).

Table 5. The effects of *cis*-Pt(II) and multiple drug combinations against L1210 leukemia in mice

Drug	Doses (mg/kg)	% Prolongation		% Survivors
		Day	of life	
cis-Pt(II)	2.5	3	25—35	0
Methotrexate	4			
Thioguanine	5	3	66	0
Cyclophosphamide	30			
cis-Pt(II)+MTX, TG, CTX	as above		200	100
cis-Pt(II)	2.5	3	25—35	0
5-Fluorouracil	6			
Methotrexate	4	3	60	0
Cyclophosphamide	30			
cis-Pt(II)+5-FU, MTX, CTX	as above		200	100
cis-Pt(II)	2.5	3	25—35	0
5-Fluorouracil	6			
Thioguanine	5	3	50	0
Cyclophosphamide	30			
cis-Pt(II)+5-FU, TG, CTX	as above		200	100

L 1210 leukemia. The combination of the platinum compound with cytosine arabinoside, 5-fluorouracil, BIC, methotrexate, phosphoramide mustard, 5-azacytidine, camptothecin Na, or 5-HP showed some degree of potentiation. When *cis*-Pt(II) was used in conjunction with BCNU, however, potentiation did not occur. In fact, BCNU was more effective when used alone.

In another series of experiments these investigators studied the combination of *cis*-Pt(II) plus cyclophosphamide or ICRF-159 against advanced L 1210 leukemia. These studies showed that the advantage of combination therapy over single-drug treatment was retained against systemic disease.

Ridgway *et al.* found that *cis*-Pt(II) combined with Cytosar, cyclophosphamide or thioguanine promoted increases in lifespan at least double that achieved with the more effective of the individual drugs against L 1210 leukemia. All animals treated with combination therapy survived until the experiment was terminated. When the drugs were used individually, there were few to no long-term survivors.

In another series of experiments, they reported the effects of *cis*-Pt(II) and multiple drug combinations. The results of these experiments are summarized in Table 5.

Svec and Kovarik investigated the effects of *cis*-Pt(II) alone and in combination with cyclophosphamide or methotrexate against three diverse animal tumor systems. Leukemia RBA-34 was refractory to treatment with *cis*-Pt(II). The combination of *cis*-Pt(II) and cyclophosphamide, however, illicited a 50% increase in mean life span (MLS). The chicken Rous Sarcoma was found to be refractory to *cis*-Pt(II) or *cis*-Pt(II) combined with methotrexate. When *cis*-Pt(II) was used alone against a solid Sarcoma RB-1 in rats, it prevented tumor growth in 20% of the animals, and yielded a 100% increase in MLS in 60% of the remaining animals.

While combination therapy appears to have merit, the investigators associated with these studies have noted that therapeutic potentiation is dependent upon certain factors: 1. the maximum effective dose of each drug, 2. the most effective treatment schedule and 3. the status of the tumor e. g. early or advanced.

Combination of Platinum Compounds with Gamma-Radiation

Wodinsky *et al.* studied the effect of combining *cis*-Pt(II) and gamma-radiation against P 388 lymphocytic leukemia inoculated either IP or IV in mice. They found therapeutic potentiation when the platinum compound was administered 4 h before, immediately before or 4 h after irradiation. Potentiation was found over a wide range of drug and radiation doses.

In these experiments the timing of drug administration did not appear to influence the potentiation of combination therapy.

Combination of Platinum Compounds with Non-specific Immune Stimulants

In addition to the reports that various drugs and physical agents can enhance the antitumor effects of *cis*-Pt(II), there is some indication that the combination of the platinum compound and non-specific immunostimulants may potentiate its antitumor effect.

Conran reported that while *cis*-Pt(II) was ineffective against S-180 in BALB/c mice, the combination of zymosan and *cis*-Pt(II) promoted a significant number of regressions. The most successful of a variety of treatment regimes is summarized in Table 6. Although there were statistically significant increases in life span of some treatment groups when compared to controls, these increases of a few days were not considered to be of clinical significance.

Svec and Kovarik demonstrated that the administration of 500 µg/kg of Poly I : C on days 0, 1, 3, 5, 7, and 9 combined with 2 mg/kg of *cis*-Pt(II) given on days 1 and 9 or 4 and 14 prevented rat Leukemia RBA-34 in 30% of their animals. The MLS of the remaining animals was increased by 100%.

The importance of host defenses in promoting tumor regressions with *cis*-Pt(II) was even further substantiated by Conran. He demonstrated that the antitumor efficacy of *cis*-Pt(II) against S-180 in Swiss mice was markedly reduced if hydro-

Table 6. Results of combined zymosan and *cis*-Pt(II) therapy against S-180 in BALB/c mice

Treatment	Dose	Day of treatment	Tumor regressions at 90 days (%) [a]	Life span of non-survivors (days) mean ± S.E. [b]
Saline	0.2 ml + 0.5 ml	1 + 8	0/86 (0%)	30 ± 5
cis-Pt(II)	7 mg/kg	8	0/84 (0%)	33 ± 5 [c, d]
Zymosan	50 mg/kg	1	2/61 (3%)	29 ± 6
Zymosan + *cis*-Pt(II)	50 mg/kg + 7 mg/kg	1 + 8	30/64 (47%)	35 ± 7 [c, d, e]

[a] Number of tumor regressions/total number of animals.
[b] S.E.-Standard error of the mean.
[c] Significant increase in life span over treatment group 1 ($P<0.001$).
[d] Significant increase in life span over treatment group 3 ($P<0.0005$).
[e] Significant increase in life span over treatment group 2 ($P<0.01$).

Table 7. Platinum compounds demonstrating antitumor activity against the ascitic form of S-180 in Swiss mice

Compound	Range (mg/kg)	Best (% ILS)	Dose (mg/kg)
cis-$Pt(NH_3)_2Cl_2$	7	30—100	7
$Pt(NH_3)_2$(mal) [b]	20— 35	95.6	35
cis-$Pt(BZam)_2Cl_2$ [a, c]	20— 240	41.8	160
cis-$Pt(pTUI)_2Cl_2$ [a, d]	10— 80	31.7	20
Pt(N,N-Me_2en)Cl_2 [e]	10— 40	60.6	10
Pt(1,3-pn)Cl_2 [f]	7— 14	50.0	10
Pt(chxn)Cl_2 [g]	10— 30	119.2	15
Pt(chxn)(mal)	20— 220	66.9	80
cis-Pt(cypentam)$_2$$Cl_2$ [h]	20— 400	55.8	160
cis-Pt(cyhexam)$_2$$Cl_2$ [i]	10—1800	71.9	50
Pt(cyhexam)$_2$(mal) [a]	10—1800	44.5	40
cis-Pt(cyhexam)$_2$$I_2$ [a]	50— 200	36.0	100
Pd(cyhexam)$_2$$Cl_2$	10— 80	80.1	80
cis-Pt(IV)(cyhexam)Cl_4	20— 120	62.3	80

[a] Positive control % ILS<30.
[b] malonate.
[c] benzylamine.
[d] p-methylaniline.
[e] N, N-dimethylethylenediamine.
[f] 1,3-diaminopropane.
[g] 1,2-diaminocyclohexane.
[h] cyclopentylamine.
[i] cyclohexylamine.

cortisone, an immunosuppressive drug, was administered 7 days before or 6 h after the platinum compound. There was a moderate reduction in therapeutic efficacy when hydrocortisone was administered 7 days after the *cis*-Pt(II).

Additional Platinum Compounds Demonstrating Antitumor Activity

While *cis*-Pt(II) has to date been the most successful platinum compound in regards to antitumor activity, there has been a continuing search for additional compounds with equal or better activity. PERESIE *et al.* synthesized a number of compounds with the following substitutions: 1. alkyl amines, 2. amines with cyclic aromatic groups, 3. chelating saturated diamines, 4. amines with ring nitrogen donors, and 5. aminocycloalkanes. All compounds within each general heading were tested against the ascitic form of S-180. A 50% or more increase in life span (ILS) over control animals was considered to be significant and indicated that the compound was an effective antitumor agent.

The most effective antitumor agents from this series are listed in Table 7. All of these compounds are presently being tested against the solid form of S-180.

References

1. BOWEN, J. R., GALE, G. R., GARDNER, W. A., JR., BONNER, W. M.: *Cis*-Dichlorodiammineplatinum(II): Suppression of Adjuvant-Induced Arthritis in Rats. Submitted to Arthritis Rheum.
2. CONRAN, P.: The Role of Host Defenses in Mediating Tumor Regressions in Mice Treated with *cis*Pt(II). To be published.
3. HEYMAN, I., FLEISCHMAN, R. W., SCHAEPPI, U., COONEY, D., DAVIS, R.: Urinary Lactic Dehydrogenase (ULDH) and Alkaline Phosphatase (UAP) as Indicators of Renal Toxicity in Rhesus Monkeys Treated with *cis*-Dichlorodiammineplatinum(II). Abstract submitted to the Colloquium Protides of the Biological Fluids, Belgium.
4. KHAN, A.: Immunosuppression with Platinum Compounds. To be published.
5. PANT, H. C., VAN CAMP, L., PERESIE, H. J., SODHI, A., ROSENBERG, B.: Serum Electrophoretic Patterns in Swiss Mice Bearing S-180 and Treated with *cis*-Pt(II). To be published.
6. PERESIE, H. J., FISCHER, R. G., JR., VAN CAMP, L., ROSENBERG, B.: Chemotherapy of Platinum Amine Complexes Against Ascites S-180 Tumors in Mice. To be published.
7. RIDGWAY, H., LAPIS, S., HALL, L. M., STEWART, D. P., HILL, J. M., SPEER, R. J.: Platinum Compounds and Their Anti-tumor Effects in Mice. To be published.
8. RYPACEK, F., CERKASOV, J., DROBNIK, J.: Note on the Effect of *cis*-Dichlorodiammineplatinum(II) on the Intestinal Epithelium of the Rat. To be published.
9. SCHAEPPI, U., HEYMAN, I. A., FLEISCHMAN, R. W., ROSENKRANZ, H., LLIEVSKI, V., PHELAN, R., COONEY, D. A., DAVIS R. D.: *Cis*-Dichlorodiammineplatinum(II) (NSC-119 875): Preclinical Toxicological Evaluation of IV Injection in Dogs, Monkeys and Mice, U. S. Government Res. Report MRI-CC02-71-19-1971 and MRI-CC05-71-45.
10. STADNICKI, S. W., FLEISCHMAN, R. W., SCHAEPPI, U., ETHIER, M.: *Cis*-Dichlorodiammineplatinum(II)-Induced Ototoxicity in Guinea Pigs. Abstract submitted to the Colloquium Protides of the Biological Fluids, Belgium.
11. SVEC, F., KOVARIK, J.: The Cytotoxic Effect of *cis*-Dichlorodiammineplatinum(II) *in vitro* and Its Potency against Experimental Animal Tumors. To be published.
12. TAYLOR, D. D., JONES, J. D., ROBINS, A. B.: Metabolism of Platinum/^{14}C/Ethylenediamminedichloride in the Rat. Biochem. Pharmac. 22, 833 (1973).
13. WODINSKY, I., SWINIARSKY, J., KENSLER, C. J., VENDETTI, J. M.: Enchanced *in vivo* Activity of Combined *Cis*-Pt. Diamminedichloride (NSC-119 875 and Gamma-Radiation Therapy. To be published.
14. WOODMAN, R. J., SIRICA, A. E., GANG, M., KLINE, I., VENDITTI, J. M.: The Enhanced Therapeutic Effect of *Cis*-platinum(II)diamminedichloride against L 1210 Leukemia when Combined with Cyclophosphamide or 1, 2-bis (3,5-Dioxopiperazine-1-yl) propane or Several other Antitumor Agents. Chemotherapy 18, 169 (1973).
15. WOLF, W., MANAKA, R. C.: Radiopharmacokinetics of ^{195m}-Pt^{m}-*cis*-Diamminedichloroplatinum(II). To be published.

The Development and Clinical Testing of New Anticancer Drugs at the National Cancer Institute — Example *cis*-Platinum (II) Diamminedichloride (NSC 119 875)

S. K. CARTER and M. GOLDSMITH

The Division of Cancer Treatment (DCT) has as one of its major functions the development and clinical testing of new anticancer drugs. The Cancer Therapy Evaluation Branch of the DCT coordinates and monitors the clinical trials on these new anticancer drugs and is responsible for liaison with the Food and Drug Administration. One of the recent drugs developed and placed into clinical trial by the DCT is *cis*-platinum(II) diamminedichloride (NSC 119 875). The purpose of this paper is to outline the major steps in the development of a new anticancer drug with emphasis on the techniques of clinical evaluation utilizing the platinum compound as the specific example.

Drug development can be divided into five stages (Table 1). In the development of anticancer agents, as for all drugs, the initial step is the acquisition of materials for evaluation. Then, these materials must be screened for activity in experimental systems, and a decision is required for or against their further movement to the expensive third step of preclinical development. Here, large amounts of the compound must be acquired; clinical formulations are made; and toxicologic and pharmacologic evaluations are performed in animals to meet the regulatory requirements for entrance into clinical trial. In clinical trial, a drug moves through the three classic phases of study in man. The final step is the introduction of the drug into widespread medical practice.

The platinum coordination compounds were initially discovered by ROSENBERG *et al.* [7] when he noted that the amplified output from an audio oscillator, delivered at 1000 cycles per second at 2 amperes between platinum electrodes immersed in a liquid culture, induced cells of *Escherichia coli* to form long filaments up to 300 times the normal length. Subsequent investigations revealed that this pattern of unbalanced growth was not due to the electrical current, per se, but to one or more long-lived chemical species of platinum released from the presumably inert electrodes in the presence of ammonium and chloride ions. Subsequent studies revealed that these released platinum salts were bactericidal ([6]) and this eventually led to testing against animal tumors. Samples of NSC 119 875, along with samples of other platinum coordination compounds, were submitted to the National Cancer Institute for screening in our experimental tumor systems, and a positive result was observed in leukemia L1210 which led the DCT to undertake development for clinical trial.

Table 1. Steps in any drug development program

1.	Acquisition of materials for evaluation
2.	Evaluation and selection of materials for further development (screening)
3.	Preclinical development — formulation, toxicology, pharmacology
4.	Clinical trial — phase I, II, and III trials
5.	Introduction into medical practice

Once a drug has been shown to be reproducibly active in one of our screening systems and shown to be a unique chemical structure, further preclinical development moves along three tracks which are closely related. One track is the pharmaceutical development which involves devising a formulation of the compound adequate for toxicologic evaluation and eventual clinical trial. This involves work on both solubility and stability. If a compound is not soluble and so has to be formulated as an oral preparation, then evidence of activity in the experimental tumor systems when administered orally must be observed. The platinum compound is soluble in water to the extent of 1 mg/ml and is formulated in 10 mg vials containing 90 mg NaCl and 100 mg of mannitol. The drug is reconstituted with 10 cc. of sterile water for injection.

A second track is the detailed study in experimental tumor systems in order to elucidate an optimal schedule for the drug administration. When *cis*-platinum(II) diamminedichloride was given a full schedule dependency evaluation in the leukemia L1210 system, the compound showed a significant degree of activity over a variety of dosage schedules when administered intraperitoneally. The drug is essentially not schedule dependent. When the drug is given orally, there is no activity. Because of this lack of schedule dependency, a wide variety of schedules have been tried in man.

The third track is the large animal toxicology schedules. After it is determined that a drug will be developed further toward clinical trial, questions must be answered regarding the highest nontoxic dose in dogs and monkeys, the nature of toxicity, the major organ toxicity in both species, predictability of toxicity, dose-response effect, manageability or reversibility of toxicity, quantitative and qualitative consistency of findings within species, and the overall adequacy of these data [1]. This information is obtained by the NCI Laboratory of Toxicology through a series of contracts with outside institutions using a standard protocol which includes the following studies:

1. Single dose (dogs),
2. Five consecutive daily treatments (dogs and monkeys),
3. Five consecutive daily treatments, 9-day rest, repeated for three treatment periods,
4. One additional schedule based on schedule dependency testing results in mice.

The NCI examined the platinum drug in beagle dogs and rhesus monkeys, and the following organs or organ systems were damaged by the agent:

Gastrointestinal Tract — Very high doses of drug produced prompt and severe emesis, anorexia, abdominal tenderness, and diarrhea in both the dog and monkey. The histopathologic correlate of these clinical signs consists of a severe enterocolitis with occasional examples of damage to the accessory glands of digestion and even to the omental fat.

Table 2. Toxicity ratings of *cis*-platinum(II) diamminedichloride (NSC 119 875) in the dog and the monkey

	Dog				Monkey	
	Single dose		qd×5		qd×5	
	mg/kg	(mg/m²)	mg/kg	(mg/m²)	mg/kg	(mg/m²)
Highest nontoxic dose (HNTD)	0.625	(13.2)	0.1875	(3.87)	0.156 (or less)	(1.94)
Toxic dose low (TDL)	1.25	(22.5)	0.375	(7.75)	0.313	(8.0)
Toxic dose high (TDH)	2.5	(47.3)	0.75	(14.9)	1.25	(15.9)
Lethal dose (LD)	5.0	(105.7)	1.5	(31.1)	2.5	(33.6)

Table 3. Phases of clinical trial with new agents

Phase I:	Clinical pharmacology Establish maximum tolerated dose at schedule(s) tested Establish toxicity parameters and determine if toxicity is predictable, treatable, and/or reversible
Phase II:	Screening for clinical activity Treat 15—30 evaluable patients with measurable disease in each of a range of "signal" tumor types Evaluate on the basis of objective response rate and characteristics of pharmacology, mechanism of action and cell cycle sensitivity
Phase III:	Trial for recommendations of general use Controlled clinical trials Combination studies

Kidney — The kidney was the principal target for the toxic actions of the platinum compound. At all but the very lowest doses studied, there were examples of alterations of renal function, pathologic urinalyses, or deranged urinary excretion of enzymes. Histologically, the metal produced acute renal tubular necrosis in dogs only at the highest doses studied however, and tubular fat depletion or no visible changes at the lower doses. Top-dose monkeys responded with nephrosis or interstitial nephritis; whereas low dose monkeys, even those whose renal function was known to be altered on chemical grounds, often exhibited no histopathology. It is possible that the derangements of serum electrolytes seen after high doses in both species derived from these renal lesions, but most likely the gastrointestinal effects also played a cooperative role.

Lymphatic System — In both dogs and monkeys, high doses of platinum very promptly destroyed circulatory lymphocytes and brought about striking lymphoid atrophy. This action was not detected at lower doses.

Hepatic — Despite the occasional elevations of transaminases seen during the monitoring of the clinical chemistry of the treated animals, no significant histopathology (apart from mild fatty changes) was produced.

Bone Marrow — Severe hypoplasia of the bone marrow was produced by high doses of the drug in both beagles and monkeys, but this effect was only poorly expressed in the peripheral blood. That is to say hematologic monitoring often failed to give evidence of the myelotoxicity of the drug.

The quantitative aspects of the toxicity of the drug are outlined in Table 2.

Before filing an investigational new drug application (IND) with the Food and Drug Administration, the preclinical toxicology is summarized to show tolerated doses in mg/kg and mg/m^2 for each species, schedule, and route. Recommendations are made for an initial clinical dose, and warnings are stated for potential toxic effects.

Traditionally, the clinical trials of any new drug are conducted in three phases. Over the years, the DCT has followed this general approach in developing a three-phase system of clinical evaluation (Table 3) which uniquely suits antineoplastic drugs developed within the Program's linear array concept [8].

Phase I studies are limited to clinical pharmacology in patients with solid tumors which are resistant to all currently available therapy. These patients should have relatively "normal" organ function and must not have received such extensive prior therapy that toxicologic evaluation would be difficult. The estimated survival of patients must be at least two months to allow complete evaluation of toxicity. Measurable or evaluable disease is not required, and a favorable clinical response (although gratifying and significant when it occurs) is not essential at this stage.

An outline of some basic phase I study principles are given in Table 4. Drug doses are given in mg/m^2 body surface area because body surface area shows better correlation to certain metabolic and excretory functions than does body weight [5]. Moreover, body surface area can be used as a common denominator for drug dosage in adults and children and for drug dosage in different animal species [2, 5, 9, 10]. The recommended starting dose for initial clinical trial is one-third of the toxic dose low in mg/m^2 of the most sensitive large animal species which, in almost every case, is either rhesus monkey or beagle. It is important to keep in mind that calculations on a mg/kg basis would *not* follow this one-third rule. For the platinum drug, this was calculated to be 2.5 mg/m^2/d $\times$ 5 or 7.5 mg/m^2 as a single dose, and clinical trials were begun at this level.

Three patients are entered at the starting dose level chosen as described above. At least one week should pass between the entry of each of the three patients on treatment to decrease the risk of simultaneous toxicity. The patients are kept on the initial dose level for an arbitrary time period, such as six weeks (shorter if toxicity occurs).

Subsequent patients are entered at higher dose levels if no dose-limiting toxicity occurs at the preceding dose level. The dose escalations are initially rapid, with increments of escalation smaller the closer one approaches the toxic range (modified Fibonacci search scheme approach). An example of such an escalation is given in Table 5. This scheme has been used successfully in two phase I studies performed by the Medical Oncology Branch ([3, 4]) and would have been safe for agents of known value such as methotrexate [5].

Two to four weeks should pass before entry of patients into the next higher dose level in order to take advantage of the experience accrued at lower dose levels, especially when delayed and potentially dangerous types of toxicity are expected. Doses should not be escalated in the same patient, as this carries a risk of cumulative toxicity and makes the prediction of a safe starting dose for phase II studies imprecise.

Table 4. Phase I study principles

1. Initial dose chosen from large animal toxicology studies.
2. Doses not escalated in the same patient.
3. Dose escalation initially rapid with increments of escalation smaller the closer one approaches the toxic range.
4. Treat three patients at each nontoxic level and treat six patients at each level showing any toxicity.
5. Antitumor response or lack thereof is not part of decision making process to move drug into phase II studies.

Table 5. Idealized modified Fibonacci search scheme approach to dose escalation in phase I study

Drug dose	Percentage increase above preceding dose level
n [a]	—
2 n	100
3.3 n	67
5 n	50
7 n	40
9 n	30—35
12 n	30—35
16 n	30—35

[a] Starting dose n (mg/m^2).

Three patients are utilized at each nontoxic dose level and six patients at subtoxic levels. More patients can be treated at drug levels with acceptable, reversible toxicity, depending on individual variability of tolerance. Past experience shows that the tolerated dose will be defined within five, plus or minus three, dose steps. Hence, between 10 and 30+ patients will be needed for exploration of the initial dose schedule.

It is usually desirable to explore more than one dose schedule of drug administration (e. g., daily doses versus widely spaced doses). The starting level for the drug dose on the second schedule is usually one to two dose steps below the level of expected toxicity, depending on the experience gained with the initial dose schedule (type of toxicity, reversibility of toxicity, variability of individual tolerance). Exploration of other dose schedules will probably require nine to 20 additional patients per schedule.

Phase I studies elucidated the maximally tolerated dose of the platinum drug to be 20—25 $mg/m^2/d \times 5$ or 75—100 mg/m^2 as a single dose. The quantitative correlations between man and other animal species on a mg/m^2 basis are given in Table 6. Toxicity in phase I studies was noted in all the organ systems predicted for by the animal systems and in one system which the animals failed to predict for. As predicted by both beagles and rhesus monkeys (Table 7), renal toxicity was observed as the most serious side effect, and damage was seen in the marrow and gastrointestinal

Table 6. Platinum, diamminedichloro-, *cis*- animal toxicology quantitative comparison mg/m²

	Highest nontoxic dose	Toxic dose low	Toxic dose high	Lethal dose
I. Single dose				
Mouse				36—40 (LD 50)
Dog	13.2	22.5	47.3	105.7
Man		50	100—120	
II. Daily×5				
Dog	3.8	7.75	14.9	31.1
Monkey	1.9	8.0	15.9	33.6
Man		1—3.5	20—25	

Table 7. Platinum, diamminedichloro-, *cis*- Speck's comparison of organ toxicity

	Renal	Marrow	Gastrointestinal	Hepatic	Neuro
Dog	+	+	+	±	—
Monkey	+	+	+	±	—
Man	+	+	+	—	+[a]

[a] 8th cranial nerve.

tract. Ototoxicity (manifesting initially as high tone hearing loss and eventually as deafness in a few patients) has been described and this was not seen in any of the animal species which showed no evidence of any kind of neurotoxic potential for the drug.

Phase II has traditionally been a screen for antitumor activity. Inherent in any screen is a clear end point of response and a critical level of activity given this response which passes the compound on to the next step. The sensitivity of this activity level is determined by many factors including the activity of established agents in the field, the capacity to study drugs at the next level, and the overall psychological expectation from the program involved and the related scientific community. This latter aspect of expectation is often what determines how critical individuals will be of a screen regardless of its intrinsic scientific validity.

The DCT has decided to use as its screen a panel of tumors which are labeled as signal tumor types. These have recently been expanded and are outlined in Table 8. Adenocarcinoma of the breast, colon, pancreas, and ovary, and bronchogenic carcinoma are used to screen against the slow growing tumors which are the most resistant to chemotherapeutic attack. The acute leukemias and lymphomas have been chosen to screen against the rapidly growing and more chemotherapeutically sensitive tumor types. It is felt that this kind of disease-oriented phase II trial is superior to the often used general phase II trial in which a wide range of tumor types are treated, but only a few of these types in a number large enough to give meaningful results. It is not uncommon in the literature to see a drug described as inactive based on a low percent of response in 50 or so patients, and to see that only two or 3 patients with colon cancer or breast cancer were treated.

The leukemias and lymphomas are, in one sense, the easiest to evaluate from a phase II point of view, as response criteria are well established and readily accessible in almost every patient. In another sense, however, they are difficult, as the patients reaching the stage of phase II trial have usually been through extensive prior chemotherapy and are at a far advanced and relatively refractory stage of their disease. Given the great potential of single agents and combinations to achieve high complete

Table 8. "Signal" tumors of the DCT's drug development program

1. Adenocarcinoma of the breast
2. Adenocarcinoma of the colon
3. Bronchogenic carcinoma
4. Adenocarcinoma of the pancreas
5. Ovarian cancer
6. Malignant melanoma
7. Acute myelocytic leukemia
8. Acute lymphocytic leukemia
9. Lymphomatous disease
10. Malignant gliomas

remission rates, a new drug attempting to establish itself faces a particularly rough hurdle. It is recognized, therefore, that other criteria besides pure percent activity have to govern whether a drug passes the phase II screen in the rapidly growing tumors. A drug without bone marrow toxicity which could be easily included into new combination regimens would obviously be superior to a traditional marrow toxic agent with a similar response rate. Drugs which can enter pharmacologic sanctuaries. such as the central nervous system, are highly desirable, as are drugs which have the ability to kill nonproliferating or G_0 cells.

In the solid tumor area the problems have a different emphasis to them. Here, complete remission by a single agent is an unfortunately rare occurrence, and partial response occurs in the range of 20—30% for most active agents. This means the sensitivity level for activity can be placed lower and patients treated at an earlier therapeutic stage in relation to drugs. The major problem in this area is the lack of a clear cut satisfactory parameter for response and the heterogeneity of the patient population for study.

While hints of antitumor activity have been observed in the initial clinical evaluations, full fledged disease-oriented phase II trials have been slow in developing. Currently, a trial is nearing completion in gastrointestinal cancer at the Mayo Clinic, and while the preliminary results are not encouraging, many patients are still on study. Broader phase II trials in other diseases are now being planned as are a series of combination studies evolving out of the experimental data in leukemia L1210 indicating that platinum is synergistic with 5-fluorouracil, Cytoxan, and other drugs.

The National Cancer Institute is not directly involved in the commercial application of drugs developed under its auspices. Marketing is within the purview of the pharmaceutical industry. Many times the marketing rights belong to the pharmaceutical company that was the original supplier of the drug. In other instances, no pharmaceutical firm is involved in the development of a successful drug, and the "orphan" compound is made available to the industry under a bidding mechanism not dissimilar to a contract. The company demonstrating the optimal ability to prepare and market the compound is given exclusive rights to the NCI clinical data required for preparing a new drug application (NDA).

It is too early to say whether this first of the platinum coordination compounds will eventually make it to the end of the drug development road as a compound of established clinical usefulness. There is no doubt that the structure has antitumor activity, and future work on analogues should be of great interest.

References

1. Dixon, R. L.: Laboratory of toxicology. Cancer Chemother. Rep. **2(3)**, 61 (1971).
2. Freireich, E. J., Gehan, E. A., Rall, D. P., Schmidt, L. H., Skipper, H. E.: Quantitative comparison of toxicity of anticancer agents in mouse, rat, hamster, dog, monkey, and man. Cancer Chemother. Rep. **50**, 219 (1966).
3. Hansen, H. H.: Clinical experience with 1-(2-chloroethyl)3-cyclohexyl-1-nitrosourea (CCNU, NSC 79 037). Proc. Amer. Ass. Cancer Res. (Abstr.) **11**, 87 (1970).
4. Muggia, F. M.: Phase I study of 4′ demethyl-epipodophyllotoxin-*B*-D-thenylidene glycoside (PTG, NSC 122 819). Proc. Amer. Ass. Cancer Res. (Abstr.) **11**, 58 (1970).
5. Pinkel, D.: The use of body surface area as a criterion of drug dosage in cancer chemotherapy. Cancer Res. **18**, 853 (1958).
6. Rosenberg, B., van Camp, L., Grimley, E. B., Thomson, A. J.: The inhibition of growth on cell division in *Escherichia coli* by different ionic species of platinum(IV) complexes. J. biol. Chem. **242**, 1347 (1967).
7. Rosenberg, B., van Camp, L., Krigas, T.: Inhibition of cell division in *Escherichia coli* by electrolysis products from a platinum electrode. Nature (Lond.) **205**, 698 (1965).
8. Rothenberg, L., Terselic, R. A.: Management of the National Cancer Institute's drug research program through application of the linear array concept. Cancer Chemother. Rep. **54**, 303 (1970).
9. Schein, P. S., Davis, R. D., Carter, S., Newman, J., Schein, D. R., Rall, D. A.: The evaluation of anticancer drugs in dogs and monkeys for the prediction of qualitative toxicities in man. Clin. Pharmacol. Ther. **11**, 3 (1970).
10. Shirkey, H. C.: Drug dosage for infants and children. J. Amer. med. Ass. **193**, 443 (1965).

Further Clinical Experience with *cis*-Platinum (II) Diamminedichloride

J. M. HILL, E. LOEB, A. S. MACLELLAN, N. O. HILL, A. KHAN, and J. KOGLER

Following the work of ROSENBERG *et al.* [1] and ROSENBERG and VAN CAMP [2] extensive trials of *cis*-Platinum Diamminedichloride (PDD) in tumor bearing animals were undertaken by SPEER *et al.* in our laboratories. Based on these animal studies clinical trials were started in April, 1971. The results observed in twelve patients receiving PDD were presented in Prague in August of the same year. Marked tumor regression was noted in several patients and observations on tissue distribution as well as pertinent information on dosage levels and schedules were obtained.

In the present series of patients the usual dose of PDD was $^1/_2$ mg/kg given intravenously twice a day at a rate not exceeding 1 mg/kg/h and for a total dose of 2 to 4 mg/kg. For the purpose of this study a course of PDD therapy is defined as one or more doses given over a period of time up to fourteen days. In patients showing a satisfactory response repeated courses of treatment were given at four to six week intervals as monitored by bone marrow recovery, and kidney function. Results were classified as less than 25% improvement, 25%—50% improvement and over 50% improvement. Improvement was graded by all available means of objective observation including direct measurement, physical examination, x-rays and scanning techniques, plus the more subjective assessment of the clinical state according to physical activity, appetite, freedom from pain, etc. Most of our patients with solid tumors were late in the disease, resistant to other therapy and many had had the primary lesion removed, most of our studies involving metastatic or direct extension lesions. As of November 1, 1972 a total of 126 patients had received PDD. An additional twenty-three patients since then have received PDD, up to April 1, making a total of 149 treated. The results in thn latter group will be mentioned as an addendum. 67 out of the 126 patients had received PDD as single agent therapy as their initial treatment on our service. The remainder of the total group received PDD in combination with other agents. Failure of prior treatments or sufficient elapse of time had occurred so that the results could reasonably be ascribed to PDD.

Fig. 1 shows the distribution of malignant diseases and the results obtained, listed in descending order of improvements. 16 patients with acute leukemia also received PDD with only one, an adult with acute myelogenous leukemia, achieving a complete but brief remission on this agent. The overall rate of improvement for malignant diseases other than leukemia showed 20% of the patients with improvement of more than 50%, 7% in the 25%—50% group and 73% poor response or failures. With the exception of the very good results with squamous cell carcinoma and pos-

Neoplasm	Number	Improvement or regression		
		<25%	25%—50%	>50%
Endometrial carcinoma	2	0%	0%	100%
Hepatoma	2	50%	0%	50%
Squamous cell carcinoma	12	50%	8%	42%
Sarcomas — Lymphomas (7) Osteogenic (2) — Leiomyosarcoma (1)	10	60%	0%	40%
Lung undifferentiated ca.	4	75%	0%	25%
Ovary	1	0%	100%	0%
Breast carcinoma	4	75%	25%	0%
Stomach + colon carcinoma	10	90%	10%	0%
Carcinoma, kidney (1) — Pancreas (2) Prostate (1) — Myeloma (2)	6	100%	0%	0%
Total	51	73%	7%	20%

Fig. 1. Single course — PDD only chemo R_x Results

sibly the lymphosarcomas, the results were somewhat disappointing as compared with earlier trials with tumor bearing animals.

Because of the good results obtained with combined therapy in animals by Speer *et al.* [3] fifty-nine patients out of our total series were given PDD in combination with other agents as their first chemotherapy on our service. Patients in both groups who obtained good results were continued on additional courses of therapy at four to six week intervals as monitored by the leukocyte and platelet counts as well as blood chemistry or kidney function. The principal combination with PDD, designated P.A.C.T. consisted of a single dose of 2 mg/kg of PDD given intravenously in a two hour period. Amethopterin 1½ mg/kg, Cytoxan 7½ mg/kg, and Thioguanine 2½ mg/kg were also given on the same day. The Thioguanine was given one hour after the PDD and the other agents were introduced sequentially by I.V. Some patiens received P.A.C.T.(II) which is identical to the above but with an increase in Cytoxan to 15 mg/kg. Some patients receiving P.A.C.T. had PDD administered on the ½ mg/kg b. i. d. schedule, the rest of the combination being given on the second day. The overall result in this second group receiving combination PDD therapy was 7% in the over 50% improvement group, 12% in the 25%—50% group and 81% failures. However, the total dose of Platinum received by this group averaged almost 1 mg/kg less than the patients receiving PDD alone in their first course of therapy.

Most of our patients received anti-emetic therapy one hour prior to the administration. Nevertheless, 75% of the patients experienced nausea and/or vomiting. Only 2% of the patients were classified as severe nausea and vomiting. Twenty-five percent of the patients reported neither symptom.

Audiometer tests were not performed. Fig. 2 shows common side effects. The chief dose limitation factor in PDD therapy is the toxic effect on the bone marrow and the kidney tubule. Fig. 3 represents the average platelet counts for 67 patients receiving a course of PDD alone. Day 1 on the graph represents the first dose of a course of

		No. of patients	% of total
WBC	<2000	15	29%
BUN	>25	19	37%
Uric acid	>8	25	48%
Creatinine	>2	18	35%
Platelets	<20 000	9	17%

Deafness (2), tinnitus (4), mental confusion (2)

Fig. 2. Side effects of PDD-solid tumors 51 patients

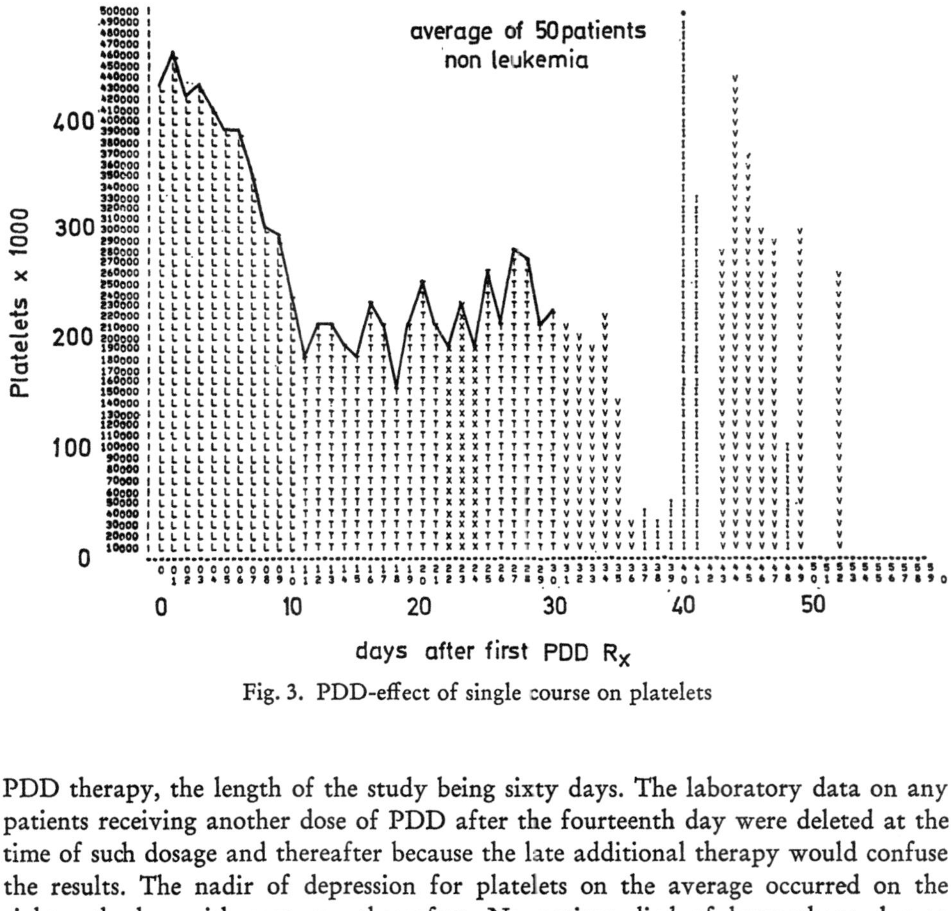

Fig. 3. PDD-effect of single course on platelets

PDD therapy, the length of the study being sixty days. The laboratory data on any patients receiving another dose of PDD after the fourteenth day were deleted at the time of such dosage and thereafter because the late additional therapy would confuse the results. The nadir of depression for platelets on the average occurred on the eighteenth day with recovery thereafter. No patient died of hemorrhage due to platelet depression resulting from PDD therapy.

Fig. 4 shows averages of leukocyte counts of the entire group of patients receiving PDD as single agent therapy but omitting patients with leukemia. In these bar graphs, in order to yield a little more information the symbols used by the computer are as follow: I — means single count, V — an average of 2—5 counts, X — 6—10 counts, T — 11—25, L — 26—50 and Y — over 50 counts per average point represented on the bar on a given day.

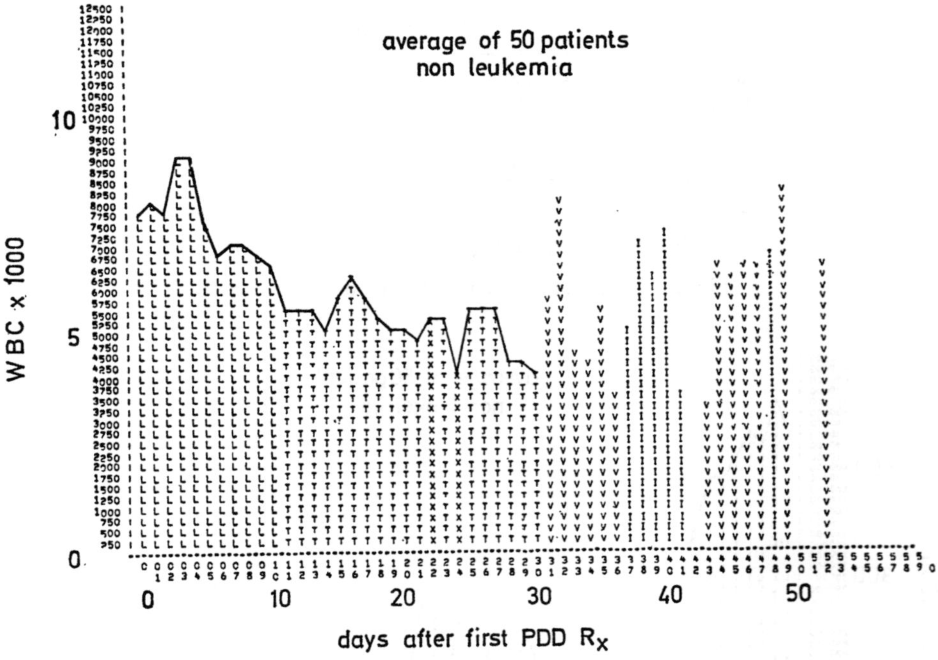

Fig. 4. PDD-effect of single course on WBC

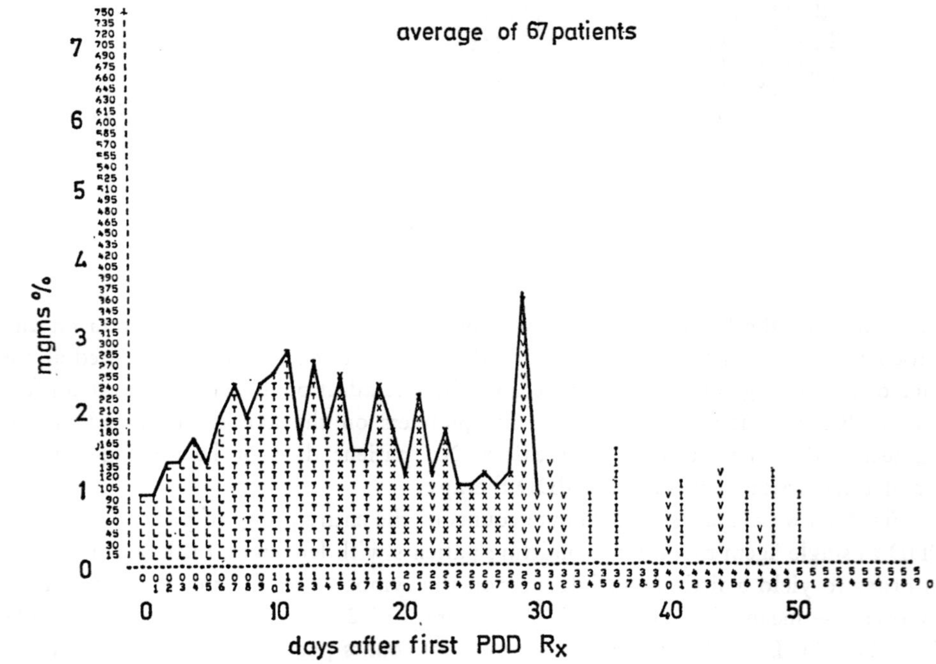

Fig. 5. PDD-effect of single course on creatinine

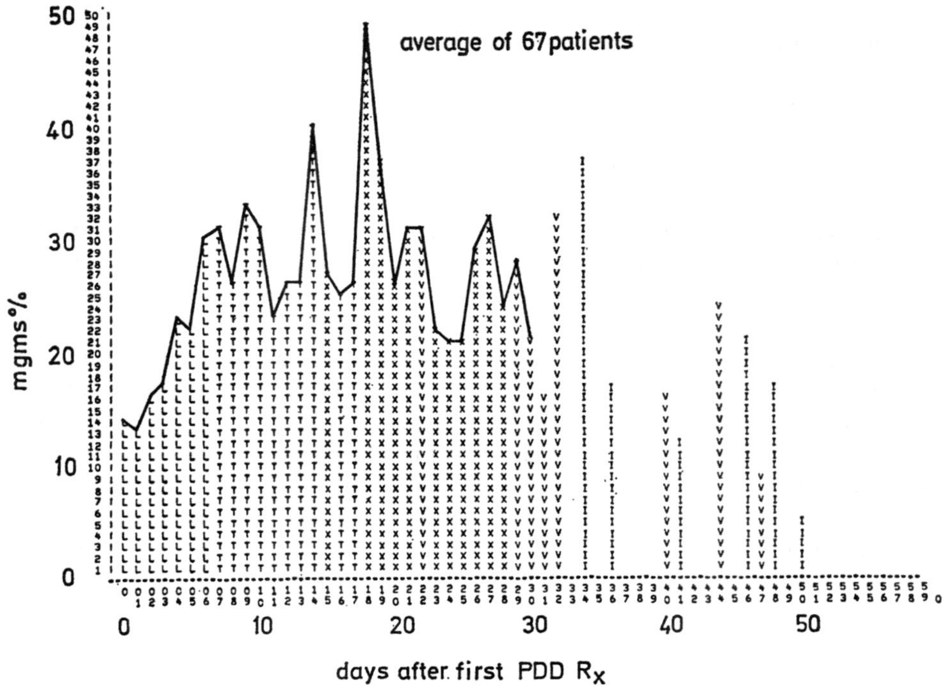

Fig. 6. PDD-effect of single course on bun

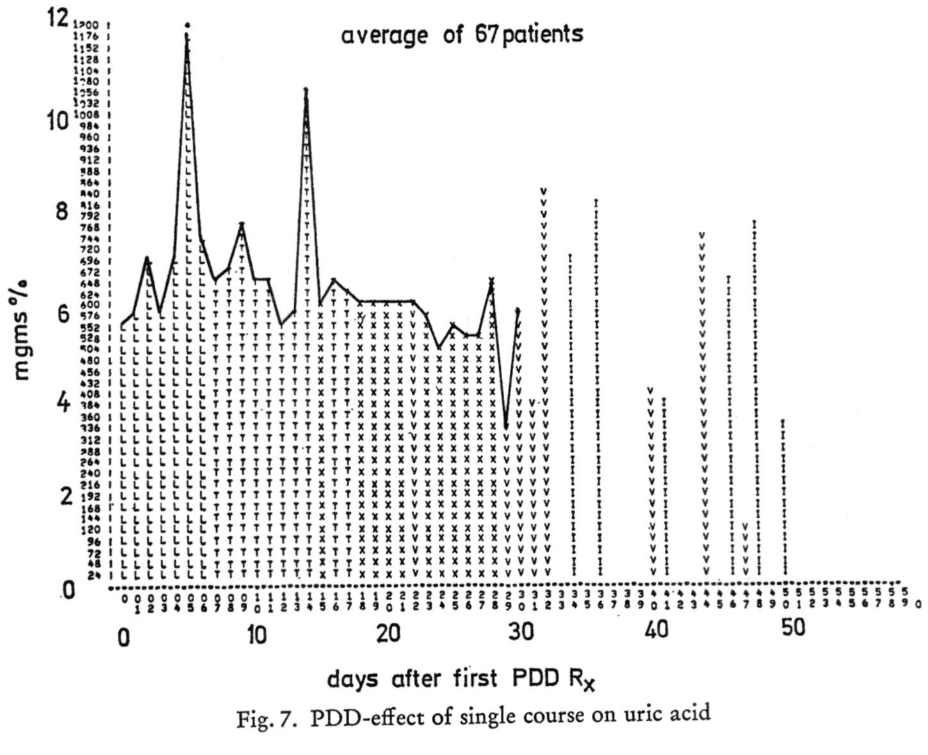

Fig. 7. PDD-effect of single course on uric acid

Fig. 5 shows the average of creatinine determinations on the sixty-three patients receiving the PDD as single agent therapy. The highest creatinine levels are observed around the eleventh and twelfth days after start of treatment.

Fig. 6 shows the average blood urea nitrogen levels after start of therapy and is similar in shape to the creatinine graph, the high point, while not sharply defined, seems to run from nine to eighteen days with diminishing levels thereafter, and with approximately normal levels after the thirty-sixth day.

Fig. 7 shows uric acid levels highest between the fifth and fourteenth days and probably relate to tumor destruction as well as to tubular injury. Severity of renal damage seemed to relate to dosage of PDD. The most severe effects were noted in the more rapidly administered single large injections of PDD in the first few patients as reported earlier. In only three instances was hemodialysis required in that group.

Because of the frequent renal tubular damage noted following PDD treatment, autopsy findings were analyzed. The first group of seven patients received PDD ten days to twenty-one days prior to death. All but two patients showed degenerative changes of tubular epithelium. These patients received 2—5 mg/kg PDD, but another group of five patients receiving 2—8 mg/kg PDD fifteen to forty weeks prior to death showed no tubular degeneration.

Case Reports

Fig. 8 illustrates the result of therapy and the dosage schedules in a 56 year old white woman who came to our Institute four months after diagnosis of squamous cell carcinoma of the esophagus, middle third. She also had squamous cell carcinoma of the larynx. 26 years earlier the patient had received radiotherapy for carcinoma of the larynx with apparent cure. She had received Thiotepa without response and had a gastrostomy tube. She had refused laryngectomy. When first seen in our clinic she was unable to speak above a whisper and could not swallow her saliva. Biopsy of the larynx at that time showed squamous cell carcinoma of the left ventricle true cord and left false cord. This patient received a total of $5^1/_2$ mg of PDD over a period of seven days, this was $1^1/_2$ mg more than was intended to be given. 13 days after the last treatment she developed severe bone marrow depression with platelets none seen and white cell count 1927. She was given symptomatic treatment with recovery of the bone marrow 22 days later. On tabular date, November 1, 1972, she had gained 31 pounds.

There was apparent complete disappearance of the carcinoma of the esophagus on x-ray twenty-four days after the PDD treatment. The laryngiogram showed marked regression of the tumor five months after the first treatment and no recurrence one year later. The patient continued to do well on repeated courses, but failed to return for 3 months and now has recurrence in the esophagus.

Another patient of interest was a 63 year old man who had had a right side nephrectomy performed for carcinoma of the kidney with extensive bone metastasis. He had also previously had a total of seven courses of combination chemotherapy other than PDD given at four to 6 week intervals. Because of the progression of bone lesions he was started on PDD and received 8 mg/kg body weight over a ten day period. This patient, previously somewhat hard of hearing, experienced some tinnitus

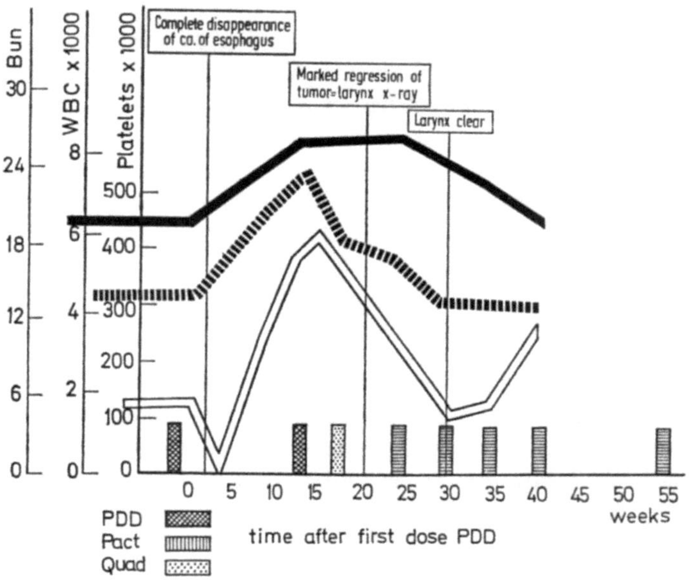

Fig. 8. Cancer of larynx and esophagus PDD and combination R_x

and some mental confusion during the therapy but did not suffer any substantial loss of hearing of the spoken voice. Prior to therapy his BUN was 12 mg.%, creatinine 9 mg.% and a creatinine clearance of 55% ml/min. 5 days after the last treatment his BUN rose to 29 mg.%, creatinine was $1^1/_2$ mg.% and creatinine clearance dropped to 29 ml/min. During the next three weeks the BUN came down to normal and his kidney function stayed within normal limits for eleven months when he expired.

Chest x-rays were taken before and after P.A.C.T. therapy in an 18 year old male with rhabdomyosarcoma. Prior to PDD therapy there was an upper mediastinal mass, and a mass around the left hilum. He had prior surgery and cobalt therapy at 2200 R to the anterior mediastinum. At the time of the P.A.C.T. treatment the liver scan showed hepatomegaly with uneven pattern of radioactivity.

Five weeks after the combined therapy with Platinum the radiologist reported "very little loss of volume of the left lung and only an appearance of fibrosis throughout the left lung, no mass at the left hilum, and almost complete disappearance of the upper mediastinal mass". The patient received five courses of P.A.C.T. but became resistant and died eleven months after the start of Platinum therapy.

A 72 year old white female referred in 1971 for treatment of a squamous cell carcinoma of the left cheek and jaw, was of interest because she had received platinum over the longest period of time, namely nine courses of P.A.C.T. therapy over a period of one year. Prior to the platinum therapy this patient had received 6750 R over the cheek and jaw without discernible improvements. There was a 3×4 centimeter fungating lesion on the inferior portion of the left cheek encroaching on the left buccal mucosa and left lower jaw, displacing the tongue medially. The x-ray of her mandible showed a small area of destruction in its anterior third but there were no cervical lymph nodes palpable. This patient had received four courses at monthly intervals of combined therapy including 5 F.U. Oncovin, Methotrexate and Cytoxan, with some improvement but the tumor was enlarging on this therapy and for this reason P.A.C.T. treatment was given. To date, sixteen months after the beginning of P.A.C.T. treatment, there has been no sign of recurrence of the carcinoma and x-ray of her mandible has shown marked improvement. Therapy has been discontinued because of a decrease in the kidney function to a creatinine clearance level of 9 ml/min, although the patient is well clinically. The patient has had no specific therapy for three months as of the date of this conference.

Addendum

In the group of twenty-three patients with various tumors that have been treated with platinum compound since November, again the squamous cell carcinomas did well with three out of six showing more than 50% improvement of the lesion and one showing partial improvement. One astrocytoma receiving 2 mg/kg PDD had a 20% decrease in tumor size on brain scan with accompanying improved alertness and activity clinically.

Conclusions

Cis-platinum(II) Diamminedichloride appears to be too good a therapeutic agent to abandon, yet too toxic for general use with the possible exception of the therapy of squamous cell carcinoma and also, but with less evidence, endometrial carcinoma. Its chief limitations are bone marrow depression, suppression of platelets and perhaps more important, kidney tubule toxicity. New and better analogs are awaited with great interest. Combinations with other therapy have given excellent results in animals but in our series of patients this has not been the case, possibly due to failure to administer sufficient quantities of the platinum component of such combination therapy. On the whole, PDD has been disappointing in leukemias. The fact that a single patient with acute myelogenous leukemia did receive a complete remission seems to justify further trials with platinum when better analogs are available.

References

1. Rosenberg, B., Van Camp, L., Trosko, J. E., Mansour, V. H.: Platinum compounds: a new class of potent anti-tumour agents. Nature (Lond.) **222**, 385 (1969).
2. Rosenberg, B., Van Camp, L.: The successful regression of large solid sarcoma 180 tumors by platinum compounds. Cancer Res. **30**, 1799 (1970).
3. Speer, R. J., Lapis, S., Ridgway, H., Meyers, T. D., Hill, J. M.: *Cis*-platinous diammino-dichloride (PDD) in combination chemotherapy of leukaemia L1210. Proceedings of the VII International Congress of Chemotherapy, Prague 1971. Special print of Adv. in Microbial and Anti-neoplastic Chemotherapy 1973.

Preliminary Report on Phase 1 Clinical Experience with Combined *cis*-diamminedichloride Platinum (II) (PDD) and 5-FU

R. A. Ellerby, F. J. Ansfield, and H. L. Davis

Early developments (National Cancer Institute Brochure, 1972; Rosenberg *et al.* [13, 14, 15]; pharmacology [1, 2, 4, 5, 7, 11, 18]) and animal tumor effectiveness ([7, 8, 12, 19, 20, 22]) have led to clinical programs [3, 16], Talley and O'Brien involving platinum derivatives. In personal communications (Schabel), discussions have centered around possible therapeutic potentiation of 5-FU by *cis*-platinum(II) diamminedichloride (PDD) in mouse L-1210 leukemia. As 5-FU was first synthesized and utilized clinically at the University of Wisconsin, much of our time has been spent in developing treatment modalities with 5-FU alone or in combination with other cytotoxic drugs. Certainly, we felt that potentiation of the tumoricidal action of 5-FU would not only benefit known responsive malignancies such as colon or breast adenocarcinomas, but could possibly induce increased activity towards minimally or previously unresponsive malignancies.

Therefore, a Phase I protocol was designed wherein patients were accepted to the study if they qualified by meeting the following criteria:

1. Presence of incurable cancer, histologically confirmed by biopsy.
2. Adequate bone marrow reserve with a white cell count greater than 4000 and platelets greater than 100 000.
3. Unimpaired renal function with serum creatinine values under 1.5 mg-% for males and 1.3 mg-% for females.
4. An acceptable oral intake without the necessity for parenteral supplementation.
5. Recovery from the toxicity of prior chemotherapy.
6. Any pre-existing infection or medical complication had to be controlled prior to the initiation of therapy.

Upon entry into the study, baseline tests were performed to delineate the extent of disease present as well as to assess the functional status of the kidneys and bone marrow. Such tests included a complete blood count, an SMA-12, urinalysis and serum creatinine, chest x-ray, bone survey and an audiogram accompanied by speech discrimination tests.

The initial design of the protocol called for monthly loading courses of PDD and 5-FU along with maintenance weekly 5-FU in between the monthly courses. 5-FU was administered by a modified multiple daily dosage (MDD) program. In this technique, the total daily dosage of 5-FU is divided into six equal amounts which are then given on an every four hour basis. Full doses are given on the first five days followed by half doses every other day for a total of four additional half doses. In our experience, when 5-FU is administered in multiple dosages, the incidence and severity of toxic reactions has been less and the total amount of drug given has been significantly more than when given by conventional single daily injections. To avoid multiple venapunctures, the drugs were injected into the heparin well of a Butterfly Infusion Set (Abbott). PDD was administered by daily injections on the first five days of the loading course.

Each patient was followed closely during the treatment with daily white cell counts and platelet counts. Every other day, urinalysis and serum creatinine values were obtained. Symptoms and signs of nausea, vomiting, stomatitis and diarrhea were recorded daily. Dosages were computed on ideal or actual weights, whichever happened to be the least.

Table 1. Comparison of MDD and conventional courses

Day	1	2	3	4	5	6	7	8	9	10	11	12	13
MDD													
5-FU	1	1	1	1	1	—	1/2	—	1/2	—	1/2	—	1/2
cis-platinum	1	1	1	1	1	—	—	—	—	—	—	—	—
Conventional													
5-FU	1	1	1	1	1	—	—						
cis-platinum	1	—	—	1	—	—	—						

Loading courses were to be terminated if the white cell count dropped below 3500, if the platelet count dropped below 100 000, if the serum creatinine value rose to twice normal values, or if the urinalysis became moderately abnormal. Also, courses were to be terminated if the patient developed three or more stools within 24 h, if stomatitis was detected or if the oral intake became inadequate.

In between loading courses, each patient was to receive weekly injections of 5-FU if his white cell count was above 3500. At the time of return for a second loading course, all baseline tests were repeated.

Dosage levels were begun at only 4 mg/kg of 5-FU and 0.125 mg/kg of PDD daily. Levels were increased by 2 mg/kg 5-FU and 0.125 mg/kg of PDD with a minimum of 2 patients at each dosage level. By the time twelve loading courses were completed, four patients had returned for a second treatment. At that point, 5-FU was at 10 mg/kg and the *cis*-platinum was at 0.5 mg/kg daily or 2.5 mg/kg/loading course. All four patients who underwent such second courses showed some degree of toxicity necessitating a change in the protocol.

A decision was made that all patients should receive only one loading course. A decision was also made to compare the toxicities of the MDD schedule with that of a more conventional course of five single daily injections. In addition, PDD was administered in two single injections, on the first and fourth day of the conventional loading course as in Table 1. The patients were selected on a non-randomized basis between the two programs, with 5-FU starting at 12 mg/kg in each. Dosages were to be increased by 2 mg/kg of 5-FU if no significant toxicity appeared in a minimum of three patients at each level. No more than 20 milliliters or 1 gram of 5-FU was given each day in either course.

The National Cancer Institute suggested that PDD administered with the new conventional course should be started at 0.5 mg/kg on each of the two days for a 1 mg/kg total. The platinum complex was increased by 0.25 mg/kg in every third patient. In the meantime, no increase was made in the platinum dosage given with the MDD technique until both courses equalled the total of 2.5 mg/kg.

At the time of discharge, all patients were begun on weekly 5-FU at 12 mg/kg. If no toxicity was observed after eight patients, 2 mg/kg increases were made. At monthly intervals, all patients returned for a recheck. If no progression occurred as determined by physical examination, x-rays or blood tests, a single PDD injection was given along with the weekly 5-FU. Monthly PDD began at 0.5 mg/kg and was increased by 0.25 mg/kg in every eighth patient.

Results

At the present time, 23 patients have received 27 loading courses. The malignancies represented in the study are in Table 2. There are 13 various cancers present, with colon, pancreas and unknown primary adenocarcinomas amounting to better

Table 2. Malignancies entered

Type	No. patients
Colon [a]	9
Pancreas or biliary	3
Adenocarcinoma, unknown primary	2
Jejunum adenocarcinoma	1
Tonsil adenocarcinoma	1
Parotid mixed cell	1
Tongue squamous cell	1
Lung squamous cell	1
Lung adenocarcinoma	1
Lung giant cell	1
Hypernephroma	1
Mesothelioma	1
Esophagus squamous cell	1

[a] One patient also has chronic lymphocytic leukemia.

than half of the 23 malignancies. 8 had failed on previous therapy. 7 of the patients in the colon or pancreas categories had advanced disease with palpable abdominal masses separate from the liver. Only two patients did not have accurately measurable lesions. 1 of the 2 had extensive spread of a squamous cell carcinoma throughout his oropharynx, neck and mediastinum while the other had a biopsy proven large abdominal mass of adenocarcinoma with somewhat obscured margins due to her short stature and 220 pound weight.

Toxicity

Table 3 outlines the variety of toxicities observed. Four courses were terminated due to the development of leukopenia. They all occurred during the MDD program. Leukopenia was evident between the sixth and the tenth days, reaching low counts of 2500 to 2800, then returning above 3000 within one to three days. A frequent observation, compatible with the findings of other observers was the development of low white cell counts two weeks after completion of therapy. Again, the resulting leukopenia was transient, not lasting more than seven days and not dropping below 2000 cells. No significant leukopenia has been noted following monthly PDD injections of up to 1.25 mg/kg.

Table 3. Toxicities observed

Type	No. patients
Hematologic	
WBC < 3500	4
Platelets < 100 000	0
Renal	
Serum creatinine > 1.5 mg-%	4
Urinary — WBC > 10/H.P.F. [a]	12
RBC > 5/H.P.F.	6
Casts > 5/H.P.F.	3
Gastrointestinal	
Nausea	24
Vomiting	20
Diarrhea	3
Stomatitis	1
Audiologic	
High tone loss	2

[a] per high power field.

Platelet counts in all patients have remained above 100 000. Over prolonged observation, in those patients who have returned for more than one maintenance phase of PDD, such counts appear to be stabilizing at between 100 000 and 200 000. No bleeding abnormalities could be attributed to the treatment program.

Nephrotoxicity was also rather mild, perhaps because only those with normal renal function were entered. Only four patients developed a serum creatinine value above 1.6 mg-%. Levels began to rise on the fifth day, reached a maximum by the 7th—9th day and then returned slowly to normal. One patient demonstrated more striking changes than the others. She had a long history of documented urinary tract infections and indeed had one during each of her two loading courses. After she had received a combined total of 3.75 mg/kg of PDD, her serum creatinine rose to 3.2 mg-% and was accompanied by white and red cells in her urine. The urinalysis abnormalities disappeared within three days, but it took two weeks for her serum creatinine to normalize.

Twelve patients developed more than ten white cells per high power field in their urinalysis. Three had urinary tract infections while a fourth had a fistulous tract between bowel, tumor and bladder leaving 8 attributable to the drugs. Five patients were noted to have greater than 5 red cells per high power field in their urinalysis. One occurred after a prostatic biopsy and another was during a urinary tract infection, leaving 3 attributable to the drugs. Greater than 5 casts per high power field were visualized transiently 3 times. No patients developed proteinuria.

Nausea was almost universal, occurring in 24 out of 27. It appeared usually 3 to 4 hours after the platinum injection and lasted between 2 and 4 h. Vomiting occurred in 20 courses, but was persistent necessitating discontinuance of the loading course in only one. Her symptoms abated on the day medication was stopped. Seven had no

Table 4. Toxicity developed during second loading courses

Patient	Toxicity	PDD total mg/kg	5-FU level mg/kg
H.K.	Leukopenia	1.75	8
D.H.	Creatinine 2.0 mg-% > 10 WBC/H.P.F. in U/A	3.125	10
G.L.	Creatinine 3.2 mg-% > 10 WBC/H.P.F. in U/A	3.75	10
F.M.	Diarrhea > 10 WBC & > 5 casts in U/A	4.375	10

vomiting while 4 vomited only once. Generally, the symptoms were mild to moderate, easily tolerated, occurring only on days of PDD injection.

Three patients developed diarrhea. All were on the MDD course, one at 10 mg/kg/day and the other 2 at 12 mg/kg/day. All had received 2.5 mg/kg of PDD. Diarrhea was brief, lasting less than 48 h. The maximum number of stools in 24 h was 4. No drugs were given to control diarrhea.

Stomatitis occurred only once at 14 mg/kg 5-FU, in a conventional course. It was mild to moderate in intensity, returning to normal in 7 days.

14 patients had bilateral high tone hearing losses prior to the initiation of therapy. 2 who had such prior hearing defects were found to have further losses at the time of follow-up audiograms. Both losses were asymptomatic. Interestingly, 3 patients had improvement in follow-up audiograms.

As mentioned previously, 4 patients received a second loading course one month after completion of the first course, and following 3 weeks of maintenance 5-FU. All demonstrated varying degrees of toxicity as outlined in Table 4.

Since the protocol change, 9 conventional courses have been completed. Present dosages are at 16 mg/kg/day of 5-FU and 2 mg/kg total of PDD. 5-FU will not be increased further, so the next patient will receive 16 mg/kg 5-FU and a total of 2.5 mg/kg *cis*-platinum. Only 1 patient demonstrated toxicity, with stomatitis on the fifth day.

In the MDD course, 5-FU has not been increased beyond 12 mg/kg as 4 of the last 6 patients have developed mild toxic reactions between the fifth and seventh days.

All received 2.5 mg/kg of PDD. Careful evaluation of further patients will be necessary to determine whether such toxicity is indeed related to the MDD technique rather than patient selection.

Although this study is being conducted to determine toxicity, 6 measurable responses have been observed.

A 55 year-old male with an anaplastic jejunal carcinoma metastatic to the right supraclavicular area had a greater than 50% reduction in measurable disease lasting 4 months.

A 60 year-old female had a tonsillar adenocarcinoma metastatic to skin and brain. She had a greater than 90% decrease in subcutaneous metastases. A sudden increase in cerebral metastases despite radiation therapy caused her death 5 months into therapy at a time when her cutaneous disease was continuing to diminish.

A 63 year-old male with squamous cell carcinoma of the lung had a decrease in his right upper lobe lesion lasting 3 months.

A 63 year-old male with chronic lymphocytic leukemia and an adenocarcinoma of the colon with mesenteric nodal and hepatic metastases has had a greater than 50% reduction in previously palpable leukemic lymph nodes and a continuing negative exam for the colon carcinoma. He is presently in his fifth month of therapy.

A 67 year-old female and a 49 year-old male both had marked reduction in abdominal masses of colonic adenocarcinoma.

Conclusion

It would appear that the mutual toxicities of 5-FU and PDD are easily tolerated when given by such loading and maintenance courses. When toxicity was significant enough to cause discontinuance of therapy, it was found to be quite transient. Responses have been brief generally, but promising, as most patients had advanced disease at the start of therapy. In review of the literature relative to animal and clinical studies, it would appear that the future of clinical usage of *cis*-platinum(II) diamminedichloride lies in combination therapy. The reports of Perlia's group earlier in this conference also point towards the advantages of such multiple drug modalities. We would be most interested in seeing the establishment of a controlled study incorporating not only 5-FU as an antimetabolite or Cytoxan as an alkylating agent, but also an antibiotic such as actinomycin D to block the rapid return of RNA and protein production as seen in platinum complex treated cells.

References

1. Gale, G., Howle, J., Walker, E., Jr.: Antitumor and antimitogenic properties of *cis*-dichloro(dipyridine)platinum(II). Cancer Res. **31**, 950 (1971).
2. Harder, H., Rosenberg, B.: Inhibitory effects of anti-tumor platinum compounds on DNA, RNA, and protein synthesis in mammalian cells *in vitro*. Int. J. Cancer **6**, 207 (1970).
3. Hill, J., Speer, R., Loeb, E., MacLellan, A., Hill, N., Khan, A.: Clinical experience with *cis*-platinum diamminedichloride, advances in antimicrobial and antineoplastic chemotherapy, Vol. 2, p. 255. München, Berlin, Wien: Urban and Schwarzenberg 1972.

4. Howle, J., Gale, G.: *Cis*-dichlorodiammineplatinum(II) — Persistent and selective inhibition of deoxyribonucleic acid synthesis *in vivo*. Biochem. Pharmacol. **19**, 2757 (1970).
5. Howle, J., Thompson, H., Stone, A., Gale, G.: *Cis*-dichlorodiammineplatinum(II). Inhibition of nucleic acid synthesis in lymphocytes stimulated with phytohemagglutinin (35 675). Proc. Soc. exp. Biol. (N. Y.) **137**, 820 (1971).
6. Kociba, R., Sleight, S.: Acute toxicologic and pathologic effects of *cis*-diamminedichloroplatinum (NSC-119 875) in the male rat. Cancer Chemother. Rep. **55**, 1 (1971).
7. Kociba, R., Sleight, S., Rosenberg, B.: Inhibition of dunning ascitic leukemia and Walker 256 carcinosarcoma with *cis*-diamminedichloroplatinum (NSC-119 875). Cancer Chemother. Rep. **64**, 325 (1970).
8. Leonard, B., Eccleston, E., Jones, D., Todd, P., Walpole, A.: Antileukemic and nephrotoxic properties of platinum compounds. Nature (Lond.) **234**, 43 (1971).
9. Mackman, S., Ramirez, G., Ansfield, F.: Results of 5-fluorouracil (NSC-19 893) given by the multiple daily dose method in disseminated breast cancer. Cancer Chemother. Rep. **51**, 483 (1967).
10. National Cancer Institute, Brochure on *cis*-platinum(II) diamminedichloride (NSC-119 875) 1972.
11. Roberts, J., Pascoe, J.: Cross-linking of complementary strands of DNA in mammalian cells by antitumor platinum compounds. Nature (Lond.) **235**, 282 (1972).
12. Rosenberg, B., van Camp, L.: The successful regression of large solid sarcoma 180 tumors by platinum compounds. Cancer Res. **30**, 1799 (1970).
13. Rosenberg, B., van Camp, L., Grimley, E., Thompson, A.: The inhibition of growth or cell division in *Escherichia coli* by different ionic species of platinum(IV) complexes. J. biol. Chem. **242**, 1347 (1967).
14. Rosenberg, B., van Camp, L., Krigas, T.: Inhibition of cell division in *Escherichia coli* by electrolysis products from a platinum electrode. Nature (Lond.) **205**, 698 (1965).
15. Rosenberg, B., van Camp, L., Trosko, J., Mansour, V.: Platinum compounds: a new class of potent antitumor agents. Nature (Lond.) **222**, 385 (1969).
16. Rossof, A., Slayton, R., Perlia, C.: Preliminary clinical experience with *cis*-diamminedichloroplatinum(II) (NSC-119 875, CACP). Cancer (Philad.) **30**, 1451 (1972).
17. Schabel, F., Jr.: Unpublished data on combination chemotherapy of NSC-119 875 with NSC-19 893.
18. Shooter, K., Howse, R., Merrifield, R., Robins, A.: The interaction of platinum(II) compounds with bacteriophages T-7 and R-17. Chem.-Biol. Interactions **5**, 289 (1972).
19. Sirica, A., Venditti, J., Kline, I.: Enhanced survival response of L-1210 leukemic mice to a single combination treatment with *cis*-platinum(II) diamminedichloride (*cis*-Pt(II); NSC-119 875) plus cyclophosphamide (CY; NSC-26 271). Proc. Amer. Ass. Cancer Res. **12**, 4 (1971).
20. Talley, R.: Chemotherapy of a mouse reticulum cell sarcoma with platinum salts. Proc. Amer. Ass. Cancer Res. 11, 78 (1970).
21. Talley, R., O'Brien, R.: Personal communication.
22. Welsch, C.: Growth inhibition of rat mammary carcinoma induced by *cis*-platinum diamminedichloride(II). J. nat. Cancer Inst. **47**, 1071 (1971).

Clinical Evalutions of Toxic Effects of *cis*-Platinum (II) Diamminedichloride

A Phase I Clinical Trial

R. W. Talley, R. M. O'Bryan, J. Gutterman, R. W. Brownlee, and K. B. McCredie

Inhibition of *Escherichia Coli (E. coli)* cell division, when exposed to electrolysis products from a platinum electrode, was first reported by Rosenberg *et al.* [3]. This same group subsequently studied several ammonium platinum salts in transplanted Sarcoma 180 ([4]) and found some of these platinum salts were capable of inhibiting tumor growth significantly in this animal model system. Later, other investigators confirmed the activity of platinum salts in a wide variety of transplanted rodent tumor models. The most consistently potent analog, *cis*-platinum(II) diamminedichloride (*cis*-Pt(II)), was selected for extensive toxicological studies in rodents, dogs, and monkeys to determine preclinical toxic effects. The highest nontoxic dose for dogs was 3.87 $mg/m^2/d \times 5$ days. Monkeys appeared to be somewhat more sensitive, with the highest nontoxic dose being 1.94 $mg/m^2/d \times 5$ days. However, the lethal dose for both dogs and monkeys was similar; 31.3 mg/m^2 for the former and 33.6 mg/m^2 for the latter. At the higher doses, the major sites involved in animal toxicological studies were kidney, gastrointestinal tract, lymphatics, and the bone marrow. As a result of these studies and reports, a Phase I clinical trial was started under the auspices of the Southwest Cancer Chemotherapy Study Group (SWCCSG) in June of 1971.

Methods

The drug was supplied by the Cancer Chemotherapy National Service Center (CCNSC) in vials containing 10 mg of *cis*-Pt(II), 90 mg of sodium chloride, and 100 mg of mannitol. The drug was reconstituted with 10 cc of sterile water, and was administered directly intravenously at the appropriate dosage. Patients were selected for this trial if their white blood count (WBC) was over 5000, platelet count was over 125 000, and BUN was less than 22 mg-%. The patient must have had no prior chemotherapy for at least four weeks, have a predicted survival of at least six weeks, and a Karnofsky performance status of 50% or better. All patients were required to sign an informed consent acknowledging they were receiving an investigational drug.

The following laboratory studies were required for all patients prior to initiation of therapy: complete blood count (CBC), platelet count, reticulocyte count, and a bone marrow aspiration; urinalysis, BUN, and creatinine or urea clearance. In addition, liver function studies such as SGOT, LDH, alkaline phosphatase, bilirubin, and BSP were required. Patients then were randomized to two dosage regimens, receiving the drug either twice weekly for three weeks or daily for five days to be repeated four weeks after the initial dose. The dose escalation schedule was established with patients on the twice weekly schedule initially receiving either 1.0, 1.5, 3.0, 5.0, 6.0, 7.5, 8.0, 10.0, 12.0, or 15.0 mg/m^2. On the five day a week schedule, the dose escalation was 1.0, 3.5, 4.0, 5.0, 6.0, 7.5, 10.0, 12.0, 15.0, 17.5, and 20 mg/m^2. On either dosage regimen, those patients receiving less than 10.0 mg/m^2, who had no significant toxic effects nor response, were escalated to a higher dose level. Randomization was carried out in the Statistical Office of the SWCCSG. During the course of drug therapy, patients were observed closely for any signs of gastrointestinal disturbance or other manifestations of clinical toxicity, and for changes in size of tumor masses. Some of the above laboratory studies were repeated twice weekly for a period of two weeks following completion of the course. In some patients, a bone marrow aspiration was repeated twenty-four hours after they received the last dose of the drug. Observations were recorded on standard SWCCSG flow sheets, and were submitted to the study coordinator for analysis. Institutions participating in this study were: Henry Ford Hospital, Detroit, Michigan; M. D. Anderson (Developmental Therapeutic Section), Houston, Texas; Lackland Air Force Base, Wilford Hall USAF Medical Center, Texas; and Baylor University, College of Medicine, Houston, Texas. The large majority of patients studied were from Henry Ford Hospital and M. D. Anderson Hospital.

As this was a Phase I clinical trial, the major objective was to determine a safe dosage regimen which would produce mild to moderate toxic effects in a majority of patients. Because of the preclinical toxicity, suggesting the kidneys, gastrointestinal tract, and hematopoietic system to be the most likely targets for insult, these systems were observed most carefully by all investigators. The following definitions of toxicity were used:

1. *Hematopoietic*
   a) Mild: Depression of WBC at levels of 2500—3500 and/or platelets 75 000 to 100 000.
   b) Moderate: Depression of WBC to levels of 1500—2500 and/or platelets to 50 000—75 000.
   c) Severe: WBC below 1500 and platelets below 50 000.
2. *Renal*
   a) Mild: Increase of BUN to 25—35 mg-% and/or creatinine to 1.5 to 2.0 mg-%.
   b) Moderate: Increase of BUN to 36—50 mg-% and/or creatinine to 2.1 to 2.5 mg-%.
   c) Severe: Increase of BUN above 50 mg-% and/or creatinine above 2.5 mg-%.
3. *Gastrointestinal*
   a) Mild: Of such a nature that concomitant medication for control of symptoms is not necessary.

b) Moderate: Require varying amounts of antiemetic drugs to control nausea and/or vomiting.
c) Severe: Symptoms not controlled by vigorous supportive medication.

Even though objective remissions were not the primary object of the study, patients were classified as having an objective remission if there was a decrease in the sum of the product of the diameters of all measurable tumor masses by 50% or greater. Decreases by 25—50% were recorded, but were not classified as objective remissions. Presence of measurable disease was not a requirement for entry of a patient on study.

Results

Toxicity

Sixty-five patients were entered on study; 57 patients were eligible for evaluation of toxicity, i. e., they survived four weeks after onset of therapy. Three patients died in less than four weeks, four patients did not complete at least one full course, and in one patient the data was insufficient for evaluation. Since one of the purposes of the project was to see if there was any possible schedule-dependency in relation to human

Table 1. Hematologic toxicity evaluation of *cis*-platinum(II) diamminedichloride

	Pt. Courses	WBC		Platelets		Hgb Fall
		2.5—3.5×10^3	<2.5	5.0—10×10^4	$<5 \times 10^4$	>2 gm
Dosage						
A. *Low-dose escalation*						
1.0— 5.1 mg/m²	11	1	0	1	0	0
6.0—12.0 mg/m²	11	1	2 [a]	0	2 [a]	1
B. 15.0 mg/m²	15	2	0	4	0	2
C. 17.5 mg/m²	5	0	3	2	1	2
D. 20.0 mg/m²	1	0	1	0	1	1

[a] on 10.0 mg/m² schedule.

toxicity, the two different schedules were selected. As can be seen from Table 1, the hematologic toxicity was definitely dose-related, and may be cumulative. The number of courses given at the lower dose levels were combined because of the relative infrequency of any significant toxicity. However, of the five patients treated with 17.5 mg/m², moderate to severe toxicity was observed in three patients. The dose of 15 mg/m² appeared to be fairly well-tolerated with only mild to moderate toxicity occurring in 15 patient courses.

Renal toxicity, as shown by increasing BUN or creatinine (Table 2), also increased in incidence with the higher dosage regimens. *Gastrointestinal toxicity,* primarily nausea and/or vomiting increased in incidence with larger doses, but the results were not as clear as those observed with renal and hematopoietic toxicity.

Table 2. Renal and G.I. toxicity evaluation of *cis*-platinum(II) diamminedichloride on 5-day schedule

	Pt. Courses	Renal						G.I.		
		Bun			Creatinine					
		25—35	35—50	50+	1.5—2.0	2.1—2.5	2.5+	1	2	3
Dosage										
A. *Low dose escalation*										
1.0— 5.1 mg/m²	10	2	1	—	—	—	—	—	—	—
6.0—12.0 mg/m²	11	1	1	1	—	1	1	3	1	—
B. 15.0 mg/m²	15	2	2	—	2	1	1	4	4	1
C. 17.5 mg/m²	5	2	—	2	—	1	2	2	2	—
D. 20.0 mg/m²	1	—	—	1	—	—	1	1	—	—

Table 3. Hematologic toxicity evaluation of *cis*-platinum(II) diamminedichloride on twice-weekly schedule

	Pt. Courses	WBC		Platelets		Hgb Fall
		$<3.5\times10^3$	<2.5	$<10\times10^4$	$<5\times10^4$	>2 gm
Dosage						
A. *Low dose escalation*						
1.0— 5.1 mg/m²	9	0	0	1	0	1
6.0—11.0 mg/m²	7	0	0	0	0	0
B. 15.0 mg/m²	14	4	2	1	0	1

The onset, nadir, and duration of leukopenia varied, beginning as early as the sixth day and as late as the 26th day after onset of therapy. The median was 14 days and, in most patients, recovery was complete by the 21st day. Thrombocytopenia tended to occur slightly later, with onset any time from the 10th to the 26th day; median for nadir was 21 days, with recovery by the 28th day in all but one patient.

Increasing BUN and/or creatinine usually was noted with 7—14 days after onset of therapy with median nadir 21 days. In patients receiving several courses of the drug, the duration of time for recovery of BUN and creatinine frequently was quite prolonged, and in three patients treatment was discontinued because of failure of these values to return to normal levels.

Toxic effects observed with the twice-weekly schedule are shown in Tables 3 and 4. Again, at lower doses hematopoietic toxicity was minimal, but some increases were observed in BUN and creatinine. In most instances, it was felt this was related to other factors such as genitourinary infection, intestinal obstruction or, in one patient, concomitant use of gentamycin. Renal toxicity occurred just as frequently at the 15 mg/m² dose level, and nausea and vomiting appeared to occur more frequently with the delayed dosage regimen as treatment progressed.

Table 4. Renal and G.I. toxicity evaluation of *cis*-platinum(II) diamminedichloride

	Pt. Courses	Renal						G.I.		
		Bun			Creatinine					
		25—35	36—50	51+	1.5—2.0	2.1—2.5	2.5+	1	2	3
Dosage										
A. *Low dose escalation*										
1.0— 5.1 mg/m²	9	1	1	2	2	1	—	1	1	0
6.0—11.0 mg/m²	7	1	1	1	1	—	1	1	—	—
B. 15.0 mg/m²	14	3	2	—	2	2	—	2	8	2

Table 5. Cumulative toxicity data on pts. receiving more than 1 course

	No.	Hematopoietic				Renal				G.I.			
15 mg/m²×5 days — each course	Pts.	0	1	2	3	0	1	2	3	0	1	2	3
1st course	4	4	—	—	—	3	1	—	—	2	1	—	1
2nd course	4	2	1	1	—	1	2	—	1	2	1	1	—
3rd course	2	—	2	—	—	0	2	—	—	1	—	1	—
15 mg/m² twice-weekly — each course	Pts.	0	1	2	3	0	1	2	3	0	1	2	3
1st course	4	3	1	—	—	—	—	—	—	1	2	1	0
2nd course	4	1	2	1	—	2	2	—	—	1	1	2	0
3rd course	1	—	1	—	—	—	—	—	—	—	—	—	1

0 — No toxicity; 1 — Mild; 2 — Moderate; 3 — Severe.

Because no real difference was observed, with the exception of gastrointestinal toxicity, the five-day schedule was used for patients receiving dosage regimens above 15.0 mg/m².

Another factor to consider in evaluation of toxicity is the possible cumulative effects of this drug on both the hematopoietic system and kidneys. *Cis*-platinum(II) undoubtedly is an alkylating agent, and it is a recognized fact that most alkylating agents have the propensity for cumulative marrow toxicity. Since the drug is partly a heavy metal, even though tightly complexed, it is possible that cumulative renal toxicity may be due to persistence of the platinum ion in the renal tubules.

Analyses of the data in patients receiving two or more courses of either regimen at a dose of 15 mg/m² are shown in Table 5. Even though only four patients received two or more courses with each regimen, there appeared to be increasing hematopoietic and possibly renal toxicity with prolonged exposure to the drug.

Other Toxicity — One patient developed marked ST-T wave changes on his electrocardiogram during therapy, and congestive heart failure required digitalization. There was no prior history of heart disease. Tinnitus was observed in one patient. Audiometric studies were done prior to therapy and 4 weeks post therapy on eight patients who received 15 mg/m^2 × 5 days or higher doses. In four patients, there was approximately a 15% loss of hearing in the higher decibel range. Two patients were metal workers and already had significant loss of high decibel hearing; this was not accentuated by the therapy. Two other patients had no demonstrable deficit following therapy.

Objective Results

Patients with a wide variety of neoplasms were treated, many at low dose levels and thus could not be considered therapeutic trials. However, of the 57 who completed one course of therapy, only five remissions were observed. The following malignant neoplasms were included: gastrointestinal 18, lung 10, genitourinary 6, lymphomas 5, head and neck 4, soft tissue sarcomas 4, unknown primaries 5, melanomas 2, and one each of adrenal, thymus, and neuroblastoma.

The best objective remission was in the patient with malignant thymoma. His therapy was initiated at 4 mg/m^2 × 5 days, and subsequently escalated to 6 mg/m^2 for his second course, and 10 mg/m^2 for his third and fourth course. Because of azotemia, the dose was reduced to 5 mg/m^2 for six more courses. The patient was maintained in remission for 13 months, and then, because of increasing disease and azotemia, therapy was discontinued.

A transient three-month decrease in hepatic metastases from a neuroblastoma was observed in a three and a half year old female. There was a 50% decrease in the size of the liver below the costal margin which lasted two months following treatment with 5 mg/m^2 and then escalated to 7.5 mg/m^2. However, prolonged azotemia prevented subsequent therapy.

Five patients with various non-Hodgkin's lymphomas were treated with doses of 15 mg/m^2 or greater; reduction in the size of palpable nodes occurred in three of these patients lasting 4 months, 3 months, and 1 month.

Even though other objective remissions were not observed, 50 patients, including those described above, had received intensive chemotherapy with one or more drugs and also, in many instances, had been treated with radiotherapy. However, only one of the patients experiencing an objective remission (the patient with the thymoma) had any subjective benefit from therapy.

Discussion

Even though *cis*-platinum(II) has produced most encouraging results in a variety of transplanted animal tumors, its use in humans may be limited by renal toxicity. However, it is possible that many dosage regimens other than those discussed here may prove beneficial. Also, adequate therapeutic trials with the dosage regimen of 15 mg/m^2 need to be explored in patients not previously treated with many chemotherapeutic agents.

Since hematopoietic toxicity does not appear to be severe at the dose level of 15 mg/m^2, perhaps combinations wtih agents having different mechanisms of action should be considered in the future. Renal toxicity, which may be cumulative, appears to be the limiting factor in prolonged utilization of this agent.

The mechanism of action of *cis*-platinum(II) is not definitely established but it has been suggested that this compound and related compounds bind DNA in a manner similar to nitrogen mustard (HARDER [1]). More recently, HOWLE *et al.* [2] reported *cis*-platinum(II) probably forms bonds with guanylate and uridylate in the nucleic acid helix, and these bonds are probably dissimilar from the binding ascribed to nitrogen mustard.

Conclusion

A phase I clinical evaluation of *cis*-platinum(II) has been underway in the SWCCSG for two years. We can conclude that a dosage schedule of 15 mg/m^2 given daily for 5 days each month is acceptable to most patients. Both the hematopoietic system and kidney function must be monitored carefully. Gastrointestinal toxicity, as manifested by nausea and/or vomiting, usually is controllable with this dosage regimen. Further exploration in more patients, and possibly evaluation of combinations of *cis*-platinum(II) with other drugs are recommended.

Acknowledgment

This study was supported in part by United States Public Health Service Grants CA-04915, CA-03392, CA-10376, and CA-03754.

The drug was furnished by Cancer Therapy Evaluation Branch, National Cancer Institute, Bethesda, Maryland.

References

1. HARDER, H. C.: Molecular effects of tumor inhibiting compounds in mammalian cells: Investigations into their possible mechanisms of action. Thesis, Michigan State University, East Lansing, 1970.
2. HOWLE, JERRY, A., GALE, GLEN, R., SMITH, ALAYNE, B.: A proposed mode of action of antitumor platinum compounds based upon studies with *cis*-dichloro-([G-^{3}H]dipyridine) platinum(II). Biochem. Pharmacol. **21**, 1465 (1972).
3. ROSENBERG, B., VAN CAMP, L., KRIGAS, T.: Inhibition of cell division in *Escherichia coli* by electrolysis products from a platinum electrode. Nature (Lond.) **205**, 698 (1965).
4. ROSENBERG, B., VAN CAMP, L., TROSKO, J. E., MANSOUR, V. H.: Platinum compounds: a new class of potent antitumor agents. Nature (Lond.) **222**, 385 (1969).

Phase I Evaluation of *cis*-Platinum (II) Diamminedichloride (PDD) and a Combination of PDD Plus Adriamycin

H. J. Wallace, Jr. and D. J. Higby

Introduction

A phase I clinical trial of *cis*-platinum (II) diamminedichloride (PDD) was initiated at Roswell Park Memorial Institute in August 1971. The pre-clinical studies of this inorganic compound indicated a striking activity in several animal tumor models including Dunning ascitic leukemia and Walker 256 carcinosarcoma, Sarcoma 180, and DMBA induced mammary tumors in female rats [3, 4, 5]. The spectrum of toxic effects in these and other pre-clinical studies suggested the possibility of some specificity of activity in human cancers. The genitourinary tract in rats, dogs, and monkeys showed evidence of severe tissue injury at less than lethal doses [1, 2]. Acute tubular necrosis with azotemia and marked atrophy of the prostate gland were seen in monkeys and dogs and testicular atrophy was seen in monkeys. We, therefore, sought patients with advanced genitourinary cancer for inclusion in this study.

Methods

The phase I study was designed to evaluate two dose schedules concurrently. Schedule I was a single large dose to be given at weekly or greater intervals for the convenience of out-patient administration. Schedule II followed the currently popular five-day course which had been used in some of the pre-clinical studies in monkeys and allows for recovery of normal tissues and immunologic responsiveness.

Patients selected for the study had measurable tumor masses of any histological types of cancer for which there was no superior treatment method available. The patients had an expected survival of at least two months and had not received any anticancer agent for at least two weeks prior to the initiation of the study. Evidence of previous myelosuppressive therapy had subsided or W.B.C. or platelet counts were stable although at a lower than normal level from disease or other factors. Renal function was normal as estimated by BUN, serum creatinine, and a creatinine clearance of greater than 70 ml/min. Patients had a baseline audiogram and EKG.

The PDD was supplied by the National Cancer Institute in 10 mg lyophilized vials. It was dissolved with 10 ml of sterile water for injection and used immediately. The drug was given by direct intravenous injection over one to five minutes.

The usual parameters were measured at least weekly and included CBC, platelets, bone marrow, 12 channel biochemical studies, amylase, serum protein electrophoresis, prothrombin time, EEG, EKG, audiogram, urinalysis, and creatinine clearance.

Results — Toxicity

Schedule I, a single intermittent dose, was initiated at 4.0 mg/m². This dose was 1/6 the single Toxic Dose Low for dogs and was to be repeated at weekly intervals if toxicity permitted. Dose escalation was to have been by modified Fibonacci search technique allowing two weeks before entry of subsequent patients at a higher dose level. Shortly after initiating the phase I evaluation of this dose schedule it was ap-

Table 1. Schedule I. Phase I evaluation of single doses of PDD — toxicity

MG/M²	Patients	Courses	Toxicity
4.0	1	1	Nausea
8.0	1	4	None
[a]			
50	3	9	Nausea and vomiting, renal, W.B.C. ↓
75	3	4	Nausea and vomiting, renal, W.B.C. ↓
100	9	13	Nausea and vomiting, renal, W.B.C. ↓, 8th nerve

[a] Study modified on basis of information from other investigators with immediate escalation to 50 mg/m².

parent that several other investigators had used single doses as high as 100 mg/m², so a direct escalation to 50 mg/m² was carried out at that time. Table 1 summarizes the experience with this dose schedule. Nausea and vomiting occurred in virtually all patients. Evidence of renal toxicity appeared in patients receiving 50 mg/m² or greater. The renal insufficiency, demonstrated by increased serum creatinine, BUN, and decreased creatinine clearances, returned to normal over a one to two months period in patients receiving 50 and 75 mg/m². At 100 mg/m², all nine patients demonstrated abnormalities of renal function after a single dose. The evidence of renal toxicity persisted for more than six months in five of the nine patients and was considered irreversible. The hematologic toxicity was mild and sporadic at all dose levels with no white count nadir less than 2000/mm³. Severe tinnitus occurred in one patient and was associated with 20% high frequency hearing loss. Another patient had asymptomatic hearing loss of 20% in the high frequency range.

The Schedule II dose was begun at 1.5 mg/m² daily times five and the courses repeated in four weeks if toxicity permitted. This dose was 1/3 the Toxic Dose Low in monkeys and was escalated in subsequent patients after the Fibonacci method. Table 2 summarizes the experience with Schedule II. Patients again complained of

Table 2. Schedule II. Phase I evaluation of five day course of PDD — toxicity

MG/M²	Patients	Courses	Toxicity
1.5	2	2	Nausea
3.0	3	3	Nausea
5.0	2	4	Nausea, W.B.C. ↓
7.5	3	5	Nausea, renal
10.0	2	6	Nausea and vomiting
12.5	2	3	Nausea and vomiting, platelets ↓
18	2	5	Nausea, renal
20	9	14	Nausea and vomiting, renal
24	3	3	Nausea and vomiting, renal, 8th nerve, W.B.C. ↓

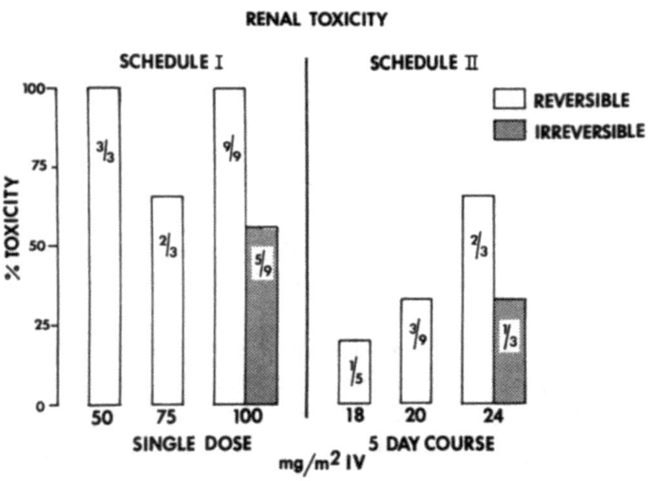

Fig. 1. Phase I PDD. Renal toxicity following one course

nausea and vomiting at all dose levels but this was ameliorated by phenothiazine antiemetics and was not dose limiting even though persistent. Renal toxicity was seen at dose levels of 18 mg/m² and above. At 24 mg/m² 2 of 3 patients demonstrated renal toxicity after a single course and the persistence in one indicated probable permanent renal damage. 1 patient had severe tinnitus with a 20% high frequency hearing loss that persisted for less than one month and did not recur when he received two more courses at a slightly lower dose level, indicating reversibility of mild 8th nerve damage.

The renal toxicity following the first course of PDD is graphically summarized in Fig. 1. The numbers of patients in each group are small but give the impression of more severe toxicity occurring at the higher dose levels when the PDD is given as a single large dose as compared to the five day course. The 5 day course in Schedule II

Table 3. Hematologic toxicity — Phase I evaluation of PDD. ( ) range

PDD MG/M^2	Leukopenia (1200—3700)	Thrombocytopenia (75 000—90 000)
Schedule I (single)		
50	0/9	2/9
75	0/4	2/4
100	0/13	2/13
Schedule II (5 day)		
7.5	1/5	0/5
12.5	1/3	0/3
20	0/9	0/9
24	0/3	1/3

Table 4. Other toxicity — Phase I evaluation of PDD

Frequent	Rare
Nausea	Hearing loss
Vomiting	Tinnitus
	Protracted anorexia
	Bloody diarrhea

appears to give a dose-response relationship that is more consistent and predictable than Schedule I. It was for this reason that the 5 day course was chosen to explore further in a phase II study.

7 of the 9 patients at the 20 mg/m² dose level eventually developed renal toxicity with repeated courses preventing further treatment. One patient underwent needle biopsy of the kidney during the acute phase of renal toxicity. Light microscopy of the specimen showed only swelling of glomeruli and tubular cells without necrosis. Unfortunately the electron microscope specimen was unsatisfactory for further delineating the pattern of renal toxicity. 2 other patients with renal toxicity died from progressive cancer following courses of PDD at 20 mg/m²/day $\times$ 5 and 100 mg/m² in a single dose. Both of these patients demonstrated histologic evidence of acute tubular necrosis. They had also received an aminoglycoside antibiotic during the terminal phase of their illness which may have been a contributing factor.

The hematologic toxicity is summarized in Table 3. Thrombocytopenia and leukopenia were sporadic, not serious, and were not dose limiting. The lowest nadir of leukopenia was 1200 and the lowest nadir of thrombocytopenia was 75 000. It was difficult to ascribe any anemia to the drug in view of the advanced nature of the patients' illness and the development of renal failure in many. One patient with severely impaired bone marrow reserve and pancytopenia following extensive radiation therapy had no further fall in hematologic parameters following institution of PDD.

Other toxic manifestations are shown in Table 4. Nausea and vomiting were almost universally present but not dose limiting. Protracted anorexia was described by one patient who had massive fibrosarcoma of the thigh involving the lower abdomen and lungs. There was only one episode of bloody diarrhea which did not recur with retreatment at the same dose level. The tinnitus and 8th nerve damage was infrequent and reversible. Unless accompanied by tinnitus, the hearing loss went unnoticed by the patients and was picked up only on the follow-up audiograms.

Results — Therapeutic Effects

Some very gratifying therapeutic effects were seen during the phase I evaluation and indicated the need to pursue further trials of this compound in spite of its rather severe toxicity. Fig. 2 shows the chest x-rays of a 42 year old female with metastatic small cell anaplastic carcinoma of the thyroid. She had received radiation therapy

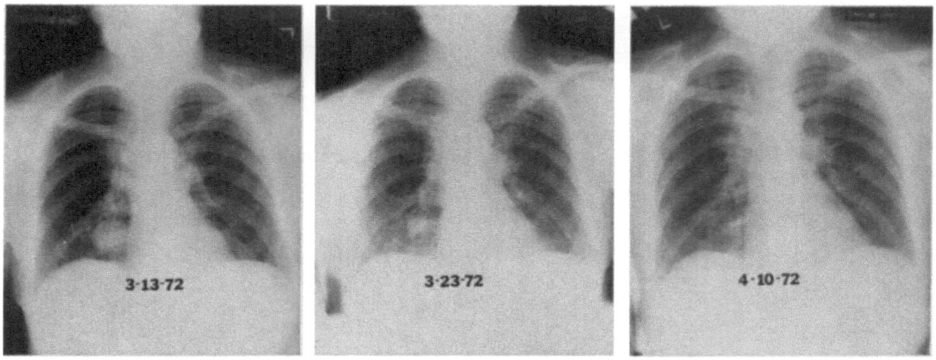

Fig. 2. Complete regression of small cell anaplastic carcinoma of the thyroid

Table 5. Testicular tumors — response to PDD by histologic type

Response	Embryonal	Choriocarcinoma	Seminoma	Mixed
CR	1/5 (5)	1/2 (2+)	1/1 (10) (3+)	—
PR	1/5 (3+)	1/2 (2+)	—	1/3 (3)
Imp.	2/5 (2) (2)	—	—	1/3 (3)
No change	1/5	—	—	1/3

Response/number of patients (months of response); CR = complete regression (100%); PR = partial regression (50—99%); Imp. = improvement (25—49%).

following surgery and presented with a right lower lung metastasis. She received two courses of PDD at 100 mg/m^2×1 and demonstrated complete resolution of the tumor associated with clinical improvement. The serum creatinine rose to 2.0 mg-% and persisted at this level for over one year preventing further treatment with PDD. The pulmonary metastases reappeared five months after treatment.

A 56 year old male had recurrent transitional cell carcinoma of the bladder with severe pain, massive hematuria, and pulmonary metastases. He received two courses

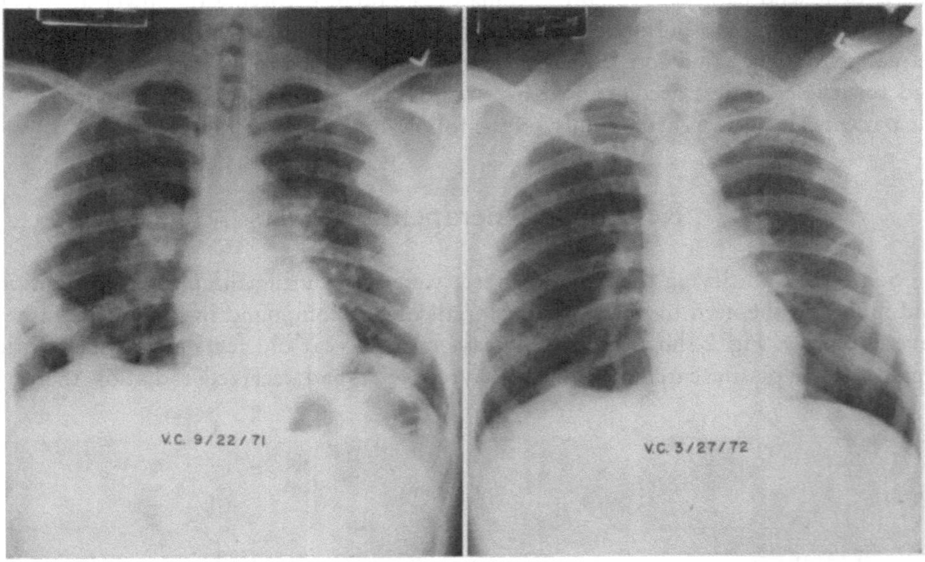

Fig. 3. Complete regression of embryonal carcinoma of testis

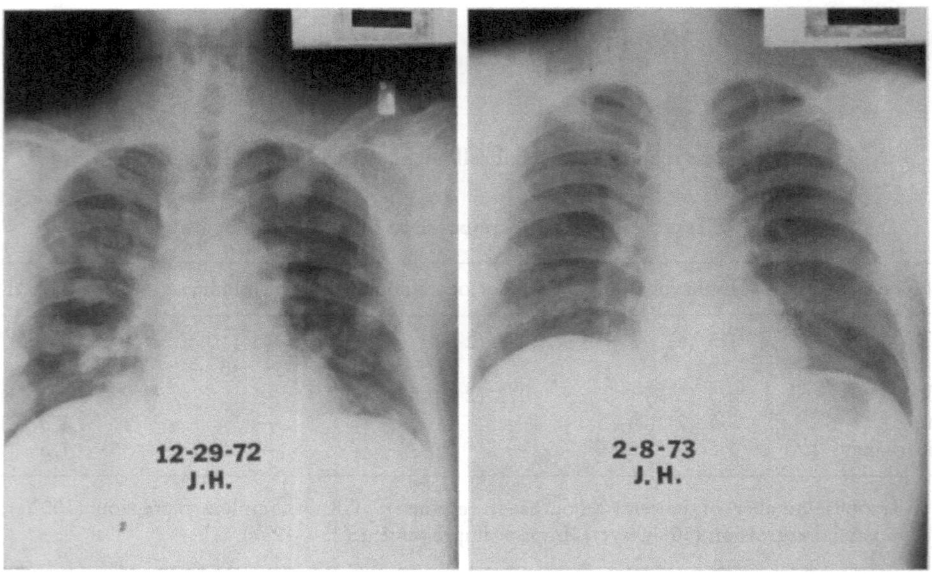

Fig. 4. Partial regression of embryonal carcinoma of testis

of PDD at a dose of 100 mg/m² until renal toxicity prevented further treatment. The patient had complete disappearance of the pulmonary metastases and the bladder tumor by cytoscopic examination plus complete pain relief and clinical improvement persisting now for more than one year.

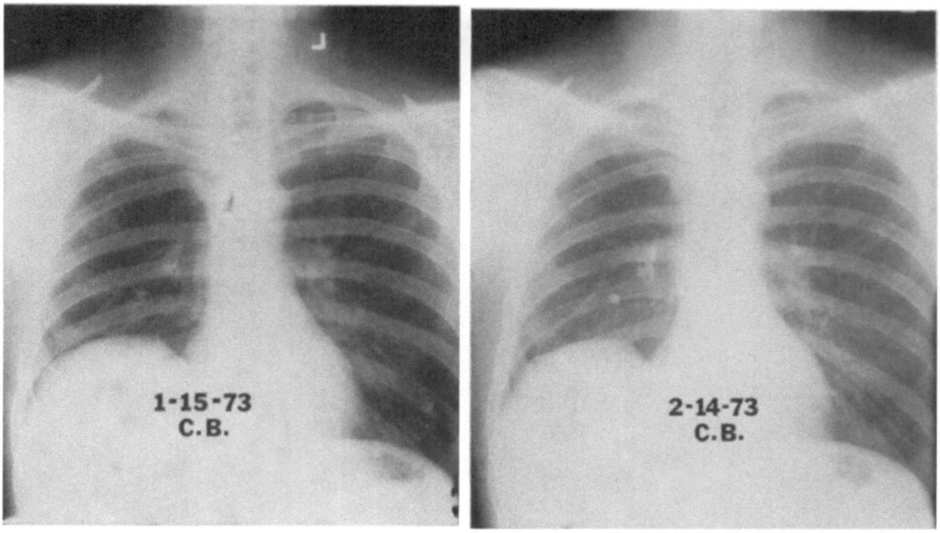

Fig. 5. Complete regression of choriocarcinoma of testis

The most striking and consistent responses have occurred in patients with testicular tumors. Table 5 summarizes the responses achieved in patients with a variety of testicular tumors.

Fig. 3 illustrates the chest x-ray of a 37 year old male with pulmonary metastases from embryonal carcinoma that had partially responded to Adriamycin and was resistant to Actinomycin D. An initial course of PDD, 24 mg/m² × 5 was given and within two weeks his symptoms had abated and the chest x-ray showed marked tumor regression. Because of striking tinnitus and 20% high frequency hearing loss which resolved in one month, he was given a second and third course at the level of 20 mg/m² × 5. Six weeks after the second course the chest x-ray was clear. The creatinine increased to 2.7 mg-% and remained elevated for greater than eight months precluding further PDD treatment in spite of tumor relapse after five months.

Fig. 4 is the chest x-ray of a 26 year old male with embryonal carcinoma treated by orchiectomy and retroperitoneal lymphadenectomy followed by adjuvant Actinomycin D. 2 months post-operatively the patient developed lung and liver metastases and he was treated with PDD, 20 mg/m² × 5. Cough and fever immediately subsided. There was rapid regression of lung lesions and decrease in liver size. Following the second course, further regression took place and he is still under treatment without apparent renal injury to date.

Fig. 5 shows the chest x-ray of a patient with choriocarcinoma of the testicle resistant to Actinomycin D. Multiple small pulmonary metastases completely regressed following two courses of 20 mg/m² × 5. The patient is continuing on treatment and the complete remission has persisted for greater than two months.

Fig. 6 illustrates the liver scans of a patient with seminoma of eight years duration treated exclusively by radiation therapy until severe radiation induced thrombocytopenia and anemia occurred. He had recurrence of cervical nodes and hepatomegaly with evidence of hepatic metastases by scan. He was given one course of Actino-

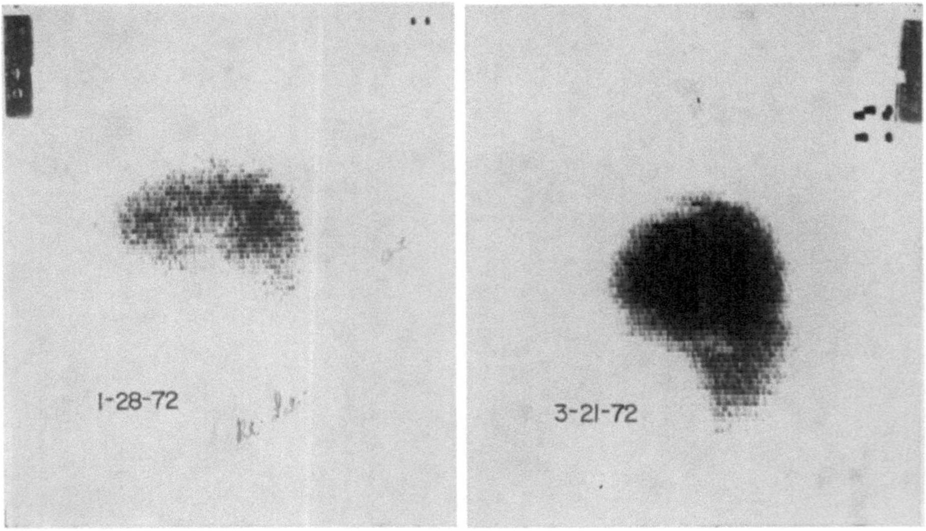

Fig. 6. Liver scans of patient with seminoma showing complete regression

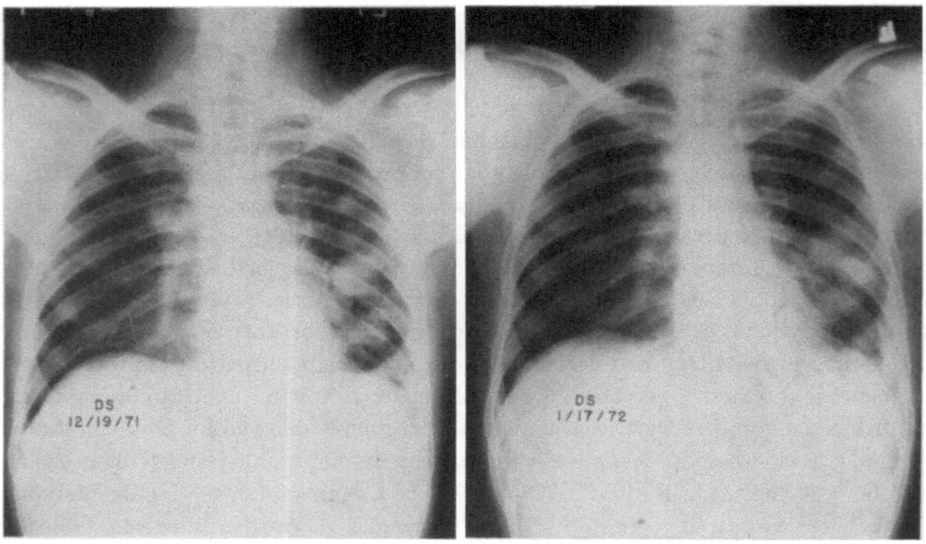

Fig. 7. Mixed response of mixed cell cancer of testis

mycin D while in a germ free protected environment unit because of the severely compromised bone marrow reserve. The patient developed serious myelosuppression but there was progression of the tumor. He then received four courses of PDD at 20 mg/m^2 $\times$ 5 with complete regression of cervical nodes following the first course and resolution of the hepatic mass following the second course. There was no further depression of the white blood cells or platelets from the drug. Elevation of serum creatinine to 2.5 mg-% precluded continuation of PDD. He remained in complete remission until an episode of herpes zoster ten months later when new tumor masses appeared in the dermatomes affected by zoster lesions. Since the renal function had returned to normal he received another course of PDD and began a prompt and complete second remission which now has lasted greater than three months.

Fig. 7 illustrates a mixed response in a patient with a mixed cell tumor. The small metastatic lesions in the right lung disappeared following a single course of PDD while the ones in the left lung showed progression.

This rather consistent evidence of antitumor activity in testicular cancers in extremely interesting and may indicate some degree of specificity for germinal cell tumors or tissues arising from the genitourinary anlage as suggested by the pre-clinical toxicity studies indicating testicular and prostatic atrophy in dogs and monkeys. Such specificity might be related to unique cell membrane characteristics. Complete response achieved in the patients with transitional cell carcinoma of the bladder is also suggestive of the G.U. anlage specificity. On the other hand, 3 patients with advanced carcinoma of the prostate and 3 patients with renal cell carcinoma failed to respond to the drug. Only one patient with advanced ovarian carcinoma has thus far been treated and she also failed to respond.

No significant responses have been demonstrated as yet in a wide variety of other advanced solid tumors although three patients with carcinoma of the breast have shown objective regression of tumors masses of less than 50%. Responses have not been seen in three patients with bronchogenic carcinoma or in the few patients with lymphoma, acute myelocytic or acute lymphocytic leukemia treated to date.

Phase II — Combined with Adriamycin

The rather encouraging response rate in previously treated patients with a variety of testicular tumors prompted an attempt to combine PDD with another drug which had demonstrated antitumor effect in these cancers. Adriamycin was chosen for the first combination as virtually all patients referred to this service had received prior

Table 6. Randomized phase II evaluation of PDD or PDD plus adriamycin

PDD 20 MG/M²/D×5	Q 4 WKS
or	
PDD 20 MG/M²/D×4	Q 4 WKS
Adriamycin 65—75 MG/M²×1	

Actinomycin D and frequently had had alkylating agents and/or an antimetabolite. A phase II study was designed as illustrated in Table 6, randomizing patients to either PDD alone or PDD plus Adriamycin. This was not intended as a comparative trial but was randomized to eliminate some bias in patient selection and to further explore the 20 mg/m² × 5 PDD dose schedule in a broad spectrum of cancers. The hematologic toxicity of the Adriamycin at 75 mg/m² was excessive in some patients with advanced cancer, many of whom had had prior myelosuppressive therapy, so it was decreased to 65 mg/m² and this dose was better tolerated.

Table 7. PDD plus adriamycin phase II — response and duration 15 patients entered

Complete regression	
Embryonal ca. testis	4 mo.
Partial regression	
Embryonal ca. testis	3 mo.
Anaplastic ca. lung	6 mo.
Sq. ca. lung	6 mo. +
Adenoca. sigmoid	3 mo.
No response	
2 renal cell	
Adrenal cortical	
Sq. ca. lung	
Sq. ca. ear	
Melanoma	
Adenoca. stomach	
Carcinoid	
Leiomyosarcoma	
Rhabdomyosarcoma	

Table 8. PDD plus adriamycin — major toxicity

	Renal (↑ bun)	W.B.C.	Platelets
Tox./no. pts.	9/15	11/15	5/15
Range	(28—145 MG⁰/₀)	(.2—2.8 $\times 10^3$/MM^3)	(5.0—97.5 $\times 10^3$/MM^3)
Median	42 MG⁰/₀	1.2 $\times 10^3$/MM^3	17.5 $\times 10^3$/MM^3

15 patients have thus far received the PDD plus Adriamycin regimen; these responses are noted in Table 7. Two patients with embryonal carcinoma of the testis responded, one with complete regression of a large retroperitoneal mass and gratifying clinical improvement lasting 4 months and another patient with partial regression of an abdominal mass lasting 3 months. Other partial responses were seen in anaplastic carcinoma of the lung, squamous cell carcinoma of the lung, and adenocarcinoma of the sigmoid colon. The types of cancer failing to show favorable response are also listed in this table.

The toxicity of the PDD plus Adriamycin combination has been summarized in Table 8 and indicates the expected instance of renal toxicity with 9 of 15 patients having BUN levels of between 28 and 145 mg-⁰/₀ with a median of 42 mg-⁰/₀. All renal function tests improved after the drugs were discontinued but many did not return to normal for several months or persisted until the death of the patient. The hematologic toxicity was excessive at the 75 mg/m^2 dose of Adriamycin with W.B.C. nadirs ranging as low as 200 cells necessitating a decrease to 65 mg/m^2. 11 of the 15 patients had leukopenia with median W.B.C. count of 1200. There was less platelet toxicity with only 5 of 15 patients having thrombocytopenia but several were severe and associated with petechiae requiring platelet transfusions. The range of platelet nadirs was from 5000 to 97 500 with a median of 17 500. The combination of PDD and Adriamycin is undergoing continuing evaluation and will be used in combination with other drugs having demonstrated activity against testicular tumors.

Studies in Progress

We are presently retesting the single PDD dose schedule at a lower level that can be given on a weekly basis for 8 weeks or longer hoping to be able to continue the drug over longer periods of time before limiting renal toxicity occurs. It appears that dose schedule juggling of this particular compound may make minor differences in the frequency and severity of renal toxicity but unfortunately will not be the answer to eliminating nephrotoxicity as a dose limiting factor. Another approach to renal toxicity is being evaluated by enhancing the renal excretion of platinum with D-penicillamine. A study is presently underway in mice utilizing D-penicillamine P.O. by gavage and PDD intraperitoneally to see if the more rapid clearing of platinum and platinum complex will modify the LD-50 dose. This is also being done in mice with leukemia L-1210 to determine if there is any adverse effect on the antitumor activity of the compound.

Another study underway is the evaluation of urinary LDH levels as an early indicator of renal toxicity. If this test proves of value for detecting early nephrotoxicity, dose schedule evaluations of new platinum coordination complex analogs will be hastened.

References

1. Heyman, I. A., Thompson, G. R., Schiappi, A. H., Rosenkrantz, H., Fleishman, R. W., *et al.*: Report MRI 6002-71-19, Contract NIH-70-2055, Dept. of HEW, PHS, NIH, NCI, Laboratory of Toxicology, 1971.
2. Kociba, R. J., Sleight, S. D.: Acute toxicologic and pathologic effects of *cis*-diamminedichloroplatinum (NSC-119 875) in the male rat. Cancer Chemother. Rep. (Part 1) **55**, 1 (1971).
3. Kociba, R. J., Sleight, S. D., Rosenberg, B.: Inhibition of dunning ascitic leukemia and Walker 256 carcinosarcoma with *cis*-diamminedichloroplatinum (NSC-119 875). Cancer Chemother. Rep. (Part 1) **54**, 5 (1970).
4. Rosenberg, B., van Camp, L.: The successful regression of large solid sarcoma 180 tumors by platinum compounds. Cancer Res. **30**, 1977 (1970).
5. Speer, R. J., Lapis, S., Ridgeway, H., Meyers, T. D., Hill, J. M.: *Cis*-platinumdiamminedichloride (PDD) in combination therapy of leukemia L-1210. Wadley Med. Bull. **1**, 103 (1971).
6. Welsch, C. W.: Effects of *cis*-platinumdiamminedichloride(II) on growth of 7,12-dimethylbenzanthracene (DMBA) induced mammary tumors in female rats. Abstract 100, Pharmacologist **12**, 2 (1970).

cis-Platinum (II) Diamminedichloride

Clinical experience of the Royal Marsden Hospital and Institute of Cancer Research, London

E. WILTSHAW and B. CARR

Between January, 1972, and March, 1973, we have been conducting a phase 2 study of *cis*-platinum (II) diamminedichloride (PDD) in the treatment of patients suffering from adenocarcinoma of the ovary. In addition we have treated 6 patients with disseminated malignant melanoma, 2 with soft tissue sarcoma and one each with lymphoma, hypernephroma and testicular teratoma. This report reviews the preliminary results in 25 patients with ovarian tumours.

Patients Admitted for Study

Patients were given PDD only when evidence of clinical resistance to other cytotoxic drugs became manifest. Of the 25 patients entered, 50% had been diagnosed more than 10 months and 25% 20 months before the platinum compound was given. All had received prior chemotherapy either with multiple drug regimes or with alkylating agents used singly. 10 had received radiotherapy to the abdomen or pelvis as well.

Treatment Regimes

The maximum tolerated dose given over 3 days was chosen for trial. The regime chosen was 30 mg/m^2 daily i.v. for 3 days and courses were repeated every 28 days. Recently 4 patients have started a regime of 30 mg/m^2 given once weekly. No more than 9 courses have been given so far to any patient but half of the patients have received more than 3 courses.

Therapeutic Benefit

In 6 of the 25 patients treated, a therapeutic effect was not assessable (3 died within a week of having PDD and 3 started the treatment less than 1 month ago). Some benefit was noted in 7 of the remaining 19 patients. The effect of the drug on tumour in various sites is given in Table 1, and shows that both pleural and peritoneal effusions responded as well as abdominal and extra-abdominal masses. Previous local radiotherapy did not prevent regression of tumour masses.

Table 1. Ovarian carcinoma effects on effusions and tumour masses

	No. cases	Response
Pleural effusions	5	2
Ascites	12	3 (+3) [a]
Extra abdominal tumour	7	2
Abdominal tumour	14	3

[a] Ascites less but always needed tapping.

Table 2. The extent of regression in 7 responding cases of adenocarcinoma of the ovary

Case no.	No. courses	Before PDD	After PDD	Length of remission
1	3	Ascites	0	3
		Abdominal mass	↓ 50%	
3	5	Nodes	0	7
		Lung deposits	↓ 75%	
		Abdominal mass	0	
9	3	Obstruction	0	3 [a]
		Pelvic mass	i.s.q.	
15	(weekly ×4)	Obstruction	0	2+
		Abdominal mass	↓ 75%	
18	2	Pleural effusions	↓	3+ [a]
		Ascites	0	
		Nodes	i.s.q.	
		Pelvic mass	i.s.q.	
20	1	Ascites	0	2 [a]
		Pleural effusions	i.s.q.	
		Nodes skin	i.s.q.	
		Pelvic mass	i.s.q.	
8	3½	Pleural effusions	↓ 75%	3
		Ascites	0	
		Breast mass	0	
		Nodes	0	

[a] Less than 50% regression or improvement.
0 = The sign disappeared. i.s.q. = the sign unchanged. ↓ = The tumour of effusion improved.

Table 2 shows the quality of response in 7 patients. Good partial regressions were obtained in 4 patients with more than 75% regression of tumour and complete or partial disappearance of ascites or pleural effusions. In 3 patients the effusions were tapped less frequently but did not disappear and/or masses regressed by less than 75%. To date, remissions have been short lived lasting 3, 3, and 7 months, and more than 2 months in one other case. All responders show some degree of improvement after the first course of treatment. For this reason patients started on therapy in the past 4 months have been reassessed after 3 courses and treatment has been abandoned if no improvement was noted at that time.

Remissions occurred in patients of all histological types and seemed to be unrelated to age or to the length of the course of their disease before treatment, but the longest remission (7 months) occurred in a patient whose tumour was diagnosed 64 months before PDD was started, perhaps suggesting a more chronic disease. The tumour was a well differentiated adenocarcinoma. Five of 19 patients have died 1, 2, 4, 5 and 8 months after starting PDD therapy while 14 patients are alive. Eleven have lived more than 4 months and 4 of these for longer than 6 months. The longest survivor is alive with recurrent tumour at 9 months.

Toxicity

The most frequent side effect was nausea and vomiting which sometimes lasted a week after the course was finished. Anti-emetics were of little help but extra fluids sometimes given intravenously for 6—12 h after each injection did reduce these unpleasant side effects which were suffered by 23 out of 25 patients.

Table 3. Side effect in 25 patients

	No.	%
Nausea and vomiting	23	92
Fall in haemoglobin by >1 g%	8	32
Rising blood urea	6	24
Tinnitus	2	8
Bundle branch block	1	4

The other frequent side effect was a falling haemoglobin (1—5 g) which occurred in 8 patients all of whom had been given 3 or more courses.

A rising blood urea has been noted after 3 or more courses of treatment in 6 patients. This has been serious in 2 cases but may be due to recurrent tumour rather than direct platinum toxicity to the kidneys. In one case a renal biopsy was performed but glomerular and tubular damage was not seen with light or electron microscopy (Table 3).

2 patients experienced transient tinnitus but in both cases audiograms were normal and treatment was continued. One other patient developed a left bundle branch block while on her second course of platinum and the drug was stopped.

Conclusions

Our experience with adenocarcinoma has shown that once resistance develops to an alkylating agent, it is extremely rare to achieve further improvement with other drugs. For this reason we are impressed that with *cis*-platinum (II) diamminedichloride we have had four definite responses to therapy and that three other patients have been marginally improved by having relief from recurrent effusions. In our opinion it has prolonged life in two instances and the remissions, in the opinion of both the patients, were worthwhile.

Summaries of the Two Best Cases

Mrs. M. B. aged 64 years

Diagnosed March, 1969. Stage IV. Serious cystadenocarcinoma of the ovary. She was given oral cyclophosphamide postoperatively and continued this treatment until July, 1970. In March, 1972 she received radiotherapy to the pelvis for recurrent pelvic tumour and in June, 1972 started platinum therapy. At that time she had multiple bilateral lung lesions and a pleural effusion, an enlarged supraclavicular node, as well as a mass arising out of the pelvis. After the first course of treatment, the lung lesions had greatly decreased in size and number and there were no other signs of disease. This improvement was maintained for 7 months but because of a falling haemoglobin (down by 5.0 g) and a rising blood urea treatment was stopped after the 5th course. Investigation of renal function including a renal biopsy failed to reveal any tubular or glomerular damage. The patient is still alive on 31. 3. 1973 but has no evidence of advancing abdominal and lung disease.

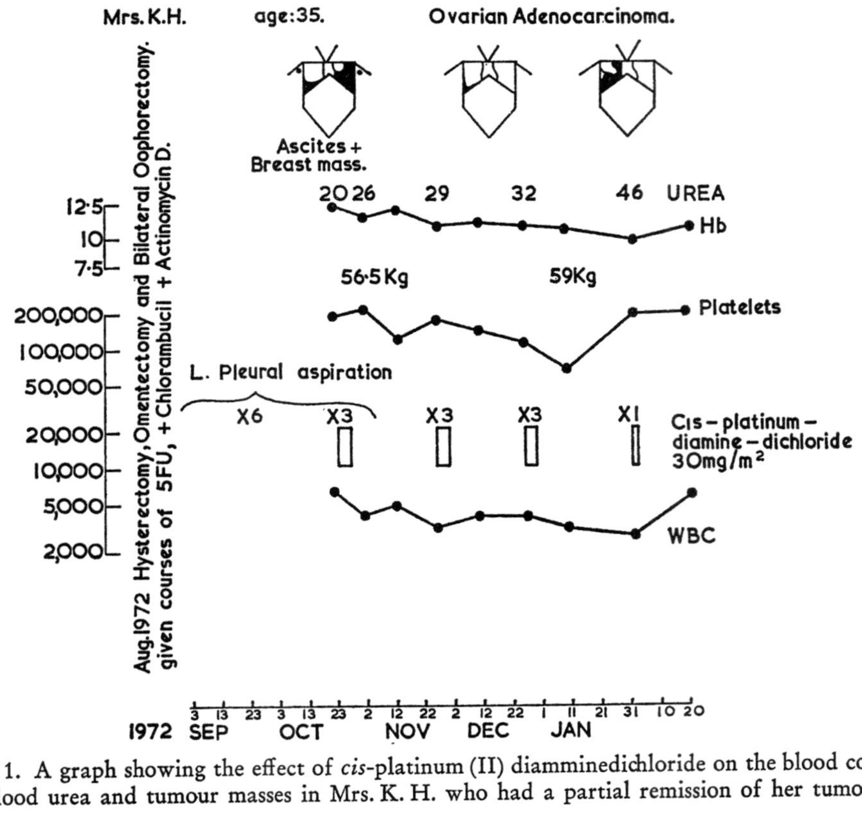

Fig. 1. A graph showing the effect of *cis*-platinum (II) diamminedichloride on the blood count, blood urea and tumour masses in Mrs. K. H. who had a partial remission of her tumour

Mrs. K. H. aged 35 years (Fig. 1)

Diagnosed August, 1972. Stage IV cystadenocarcinoma of ovary. Treatment with combinations of drugs using 5-fluorouracil, chlorambucil and Actinomycin-D were given but the tumour grew rapidly. By the end of September, 1972 she had large bilateral pleural effusions requiring tapping every 4—8 days, enlarged nodes in both axillae and a tumour mass in the lower, inner aspect of the right breast. There was gross ascites. PDD was started in October, 1972. Following the first course the nodes, breast mass and ascites disappeared. The chest was tapped on two further occasions and then no further tap was required for 4 months. The effusion in one lung never quite disappeared but it was 3 months before progressive disease occurred again. At that time she developed a mass in the right chest involving the lung and diaphragm. The axillary nodes and breast mass have not recurred so far and there is no evidence of a left pleural effusion or of ascites. She is still alive 31. 3. 1973.

Acknowledgement

I would like to thank my colleagues Drs. BOESEN, BEGENT and MCELWAIN for all their help in treating the patients included in this study: and I am most grateful to the National Institutes of Health, Bethesda, USA, for supplying the drug for this study.

Clinical Trials of *cis*-Platinum (II) Diammine-dichloride (PDD) in Patients with Advanced Cancer

I. H. Krakoff and A. J. Lippman

Cis-platinum (II) diamminedichloride (PDD) is one of a group of transitional metal compounds demonstrated to have antibiotic and antineoplastic effects *in vitro* [12, 13] and in animal tumor systems [8, 11, 14, 16]. The mechanism of action is discussed in other presentations in this symposium. It appears to involve selective inhibition of DNA synthesis [2, 4, 5, 6] with evidence for inactivation of DNA by cross-linking.

Studies in animals revealed toxic side effects involving several organ systems including the bone marrow, kidney and gastrointestinal tract [1, 15, 17]. Early clinical experience also suggested ototoxicity [3, 9]. At Memorial Hospital PDD was administered to 47 patients; the clinical data derived from this study furnish the basis of this report.

Materials and Methods

Patient Selection

32 patients with a variety of malignant tumors received PDD alone and 15 others received it together with Bleomycin. All patients had unresectable metastatic cancer proved by histologic examination, and all had measurable areas of known cancer to serve as objective criteria of response. Patients were given sufficient drug to produce toxic side effects or a clinical response.

12 patients received single courses of therapy, 5 received 2 separate courses, 3 received 3 courses, and 1 patient 4 courses. Thus, a total of 35 individual courses of therapy with PDD alone are evaluable for purposes of toxicologic study.

Methods

All patients were treated in the hospital under close observation. Daily total fluid intake and output were recorded and parenteral supplementation given if necessary to preclude clinical dehydration. The following parameters were followed: daily leukocyte and platelet counts, urinalysis, and BUN, creatinine, and uric acid levels; semi-weekly serum calcium, phosphorus, electrolytes, total protein, albumin, SGOT,

LDH, alkaline phosphatase and 5′-nucleotidase. Appropriate studies were obtained on all patients showing evidence of renal functional impairment based upon rise in BUN and creatinine.

Audiograms using a portable Beltone model 12 D audiometer were obtained on all patients prior to therapy and at weekly or biweekly intervals thereafter. All subjects were examined in their hospital beds or rooms. Background noise interfered with accurate bone conduction, which was not pursued serially. The individual threshold was determined with descending intensities on alternate ears for each frequency. The frequencies were tested in the following sequence: 4000, 1000, 8000, 500, 250, 2000, 8000, and again 1000 cps as an internal control.

Prior to commencement of therapy, renal function was determined to be normal in all patients on the basis of intravenous urogram, urinalysis, and creatinine clearance.

The planned dose of PDD was dissolved in physiologic saline at a concentration of 1 mg/ml and administered by intravenous infusion over 5—10 min. In the first 2 patients treated, the initial dose was calculated to represent one-tenth of the minimal toxic dose in mice (0.03 mg/kg/day) and increased by graduated increments at four day intervals until toxic side effects appeared [1]. Subsequent patients received various daily doses between 0.25—2.0 mg/kg for total course doses of 1.25—7.0 mg/kg.

14 additional patients were treated with Bleomycin concurrently. Those patients received a single dose of PDD, 2.0 mg/kg and Bleomycin 0.25 mg/kg daily until the development of muco-cutaneous toxicity.

Treatment was discontinued in any patient who failed to show a therapeutic response in the face of toxic side effects. 9 patients received more than 1 course of therapy. 5 of these showed no definite toxicity initially; 4 who showed a therapeutic response after the first course of therapy received one or more additional courses at a time when recovery from toxic side effects was apparent and their disease showed objective progression. All patients eventually manifested some form of toxicity.

Further administration of PDD was suspended in any patient who showed incomplete recovery of renal function, myelosuppression, hearing loss, or intolerable nausea and vomiting, even if a therapeutic benefit was obtained.

Results

Toxicity

Clinically significant toxic side effects included myelosuppression, renal functional impairment, high frequency hearing loss, nausea and vomiting.

Stomatitis, dermatitis, cystitis, and diarrhea were not observed nor did cardiac, pulmonary, central nervous system or allergic reactions occur.

Hematologic Observations

For purposes of analysis, the 35 individual courses of therapy were divided retrospectively into 2 groups, consisting of "high" and "low" total dose per course. 16 pa-

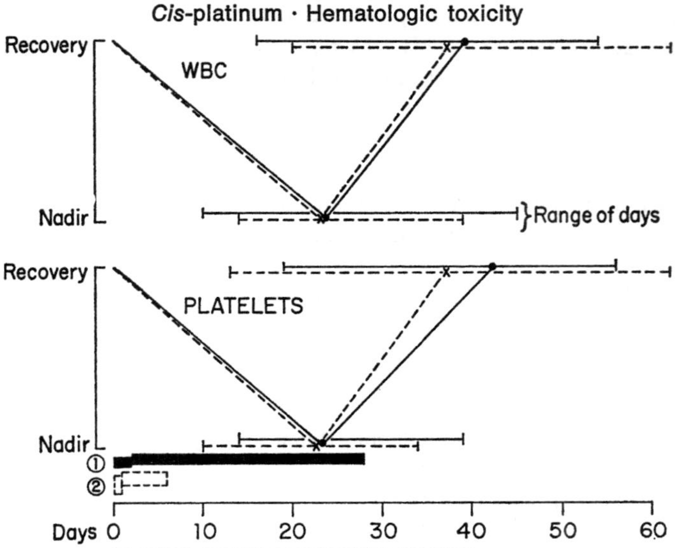

Schedule	Number Courses	Dose (mg/kg) Total	Dose (mg/kg) Daily	Duration of Rx (days)
① •——•	16	3.0-7.0	0.25-1.0	2-28
② x----x	19	1.25-2.5	0.25-2.0	1-6

Fig. 1. Time of onset and recovery from leukopenia and thrombopenia induced by two dose ranges of PDD

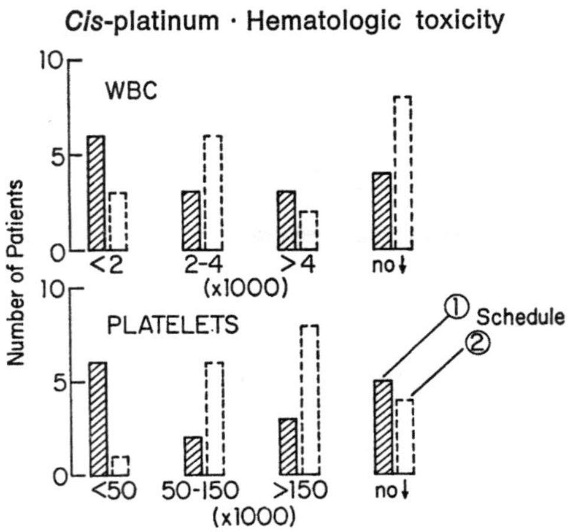

Fig. 2. Levels of leukopenia and thrombopenia in patients given two dose ranges of PDD

tients received a total course dose of 3.0—7.0 mg/kg and constitute the "high" group; 19 patients received a total course dose of 1.25—2.5 mg/kg and constitute the "low" group.

Both schedules produced leukopenia and thrombopenia and, as shown in Fig. 1, there was no apparent difference in time of onset of peripheral cytopenia. However, the higher dose schedule produced, as might have been expected, more marked degrees of myelosuppression (Fig. 2). PDD produced decrease in leukocyte counts in 2 of 4 patients with acute leukemia. This was not associated with improvement in their bone marrows nor in other manifestations of disease.

Renal Function Observations

Data pertaining to renal function evaluation are seen in Figs. 3 and 4. Again, for purposes of analysis, data were examined for both "high" and "low" dose groups as defined above.

In the high dose group all patients showed at least a transient rise in BUN. All but 2 showed a corresponding rise in creatinine. In 2 patients (patients 7 and 14) BUN and creatinine rose following therapy with PDD and continued to rise until

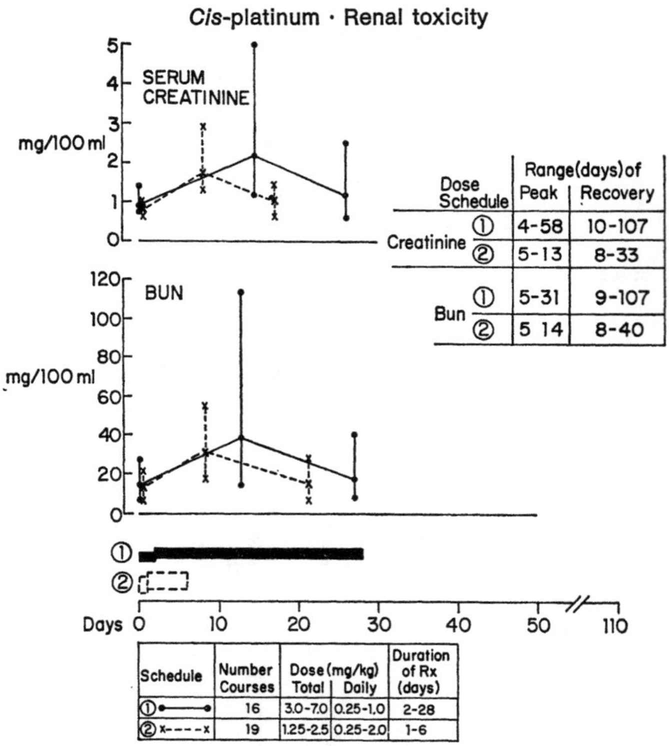

Dose Schedule		Range (days) of Peak	Range (days) of Recovery
Creatinine	①	4-58	10-107
	②	5-13	8-33
Bun	①	5-31	9-107
	②	5 14	8-40

Schedule	Number Courses	Dose (mg/kg) Total	Dose (mg/kg) Daily	Duration of Rx (days)
① •——•	16	3.0-7.0	0.25-1.0	2-28
② x----x	19	1.25-2.5	0.25-2.0	1-6

Fig. 3. Time of onset and recovery from creatininemia and azotemia induced by two dose ranges of PDD

the time of death. Both of these patients however suffered from acute leukemia, had generalized sepsis and metabolic imbalances, and were being treated concurrently with known nephrotoxic antibiotics (Gentamicin and Amphotericin B). Therefore, the role of PDD in the genesis of their renal failure cannot be assessed.

In the low dose group, 10 of 19 patients showed no change in BUN or creatinine. 2 patients showed persistent elevations of each, and all other showed only transient elevations with complete recovery to baseline values.

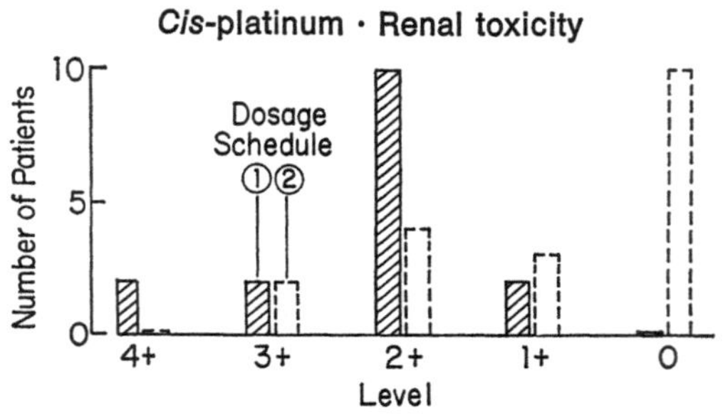

Level	Changes in: Bun	Creatinine
4+	fatal	
3+	↑ > 2x baseline, remained ↑	↑ > baseline, remained ↑
2+	Both ↑ > baseline, returned to normal	
1+	Slight transient ↑	none
0	← none →	

Fig. 4. Levels of creatininemia and azotemia induced by two dose ranges of PDD

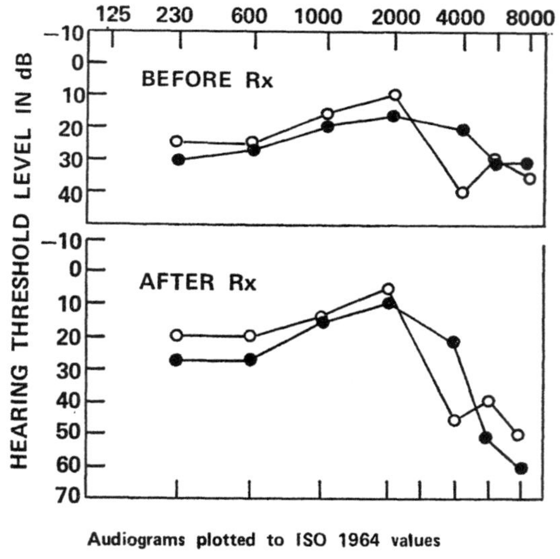

Fig. 5. Effect of PDD on high frequency hearing in a patient who received a single course of 3 mg/kg

In the 2 patients who died in uremia histopathologic examination of the kidneys revealed tubular necrosis in each. Seven other patients who died were subjected to autopsy; renal histopathological examination by light microscopy revealed a variety of lesions explainable by mechanisms related or unrelated to their basic disease.

No significant differences in toxicity were found by varying the treatment schedules. Of patients receiving more than one course only one manifested evidence of cumulative toxicity with decrease in creatinine clearance.

Audiologic Data

Apparently irreversible, high frequency (4000—8000 cps) bilateral hearing loss, presumably neural, occurred in 8 patients of the 19 studied and is demonstrated in Fig. 5.

Only one patient experienced clinically apparent hearing loss with a 40—80 db loss in the speech range (1000—4000 cps). That patient, the first treated, had received daily injections in increments up to 1.0 mg/kg on 28 consecutive days before hearing loss became apparent. No other patient experienced hearing loss in the speech range.

Cis-platinum · Ototoxicity

CUMULATIVE TOTAL DOSE mg/kg	No. PTS	DEGREE OTO-TOXICITY
2.0-3.9	4	0,0,0,0
4.0-5.9	6	0,0,1^+, 2^+,2^+,3^+
6.0-7.9	4	0,1^+,2^+,4^+
8.0-8.5	2	0,3^+

	db loss
0	–
1+	10-20
2+	20-40
3+	40-80
4+*	40-80

*clinically apparent

Fig. 6. Levels of ototoxicity induced by various dose ranges of PDD

The results are summarized in Fig. 6. Hearing loss occurred only at cumulative total doses exceeding 4.0 mg/kg. In patients who received more than one course of therapy, hearing loss appeared to be additive, becoming more severe with each course. On no occasion did audiograms clearly improve once hearing loss became established. 1 patient, who received a total dose of 4.0 mg/kg in 8 days, survived 3 months following his last dose of PDD. His final audiogram, taken 2 weeks prior to his death, suggest *slight* improvement in the high frequency range.

Gastrointestinal Reactions

Mild nausea and vomiting, occurring 2—48 h following drug administration, was seen with most courses of treatment and was easily controlled with mild antiemetics.

Therapeutic Response

Eight patients given PDD alone experienced some therapeutic benefit (Fig. 7). Responses were graded according to the standard criteria of Karnofsky [7].

All of the responses except one occurred in patients with epidermoid carcinomas of the head and neck region, and all were of brief duration. 2 patients showed I—A responses. 1 patient showed "stabilization" of his disease for 6 months and then further progression. 3 patients showed objective responses of less than four weeks duration.

One patient with adenocarcinoma of the prostate experienced subjective partial relief of skeletal pain for 8 weeks but did not manifest any objective improvement.

No other therapeutically useful responses in a variety of solid tumors and leukemias were observed.

Fourteen patients received PDD with Bleomycin. This combination was chosen because Bleomycin, too, has been found to produce some responses in squamous cancers. Three minor responses have been seen in that group.

Cis-platinum · Results of clinical trials

Diagnosis	Total	Not Yet Eval.	Ineval-uable	Eval-uable	CATEGORIES OF RESPONSE* I-A I-B I-C	II	0-C	0-B	0-A	0-0
Acute leukemia	5	0	1	4	–	–	–	–	–	4
HD + RCS	2	1	1	0	–	–	–	–	–	–
Soft part Sa's	4	0	1	3	0	0	0	0	0	3
Melanoma	1	0	0	1	0	0	0	0	0	1
Neuroblastoma	1	0	0	1	0	0	0	0	0	1
Ca Head & Neck	13	0	1	12	2(1,2)*	1(6)	2	1($^1/_2$)	0	6
Ca Prostate	1	0	0	1	0	0	0	0	1(2)	0
Ca Penis	1	0	0	1	1(pending)	0	0	0	0	0
Misc. Tumors	8	0	2	6	0	0	0	0	0	6
TOTALS	36	1	6	29	3	1	2	1	1	21
+ BLEOMYCIN										
Ca Head & Neck	12	1	3	8	2(2^+ , 3^+)	0	0	1($^1/_2$)	0	5
Misc. Tumors	2	0	0	2	0	0	0	0	0	2
TOTALS	14	1	3	10	2	0	0	1	0	7

*Durations of responses expressed in months

Fig. 7. Therapeutic results of Clinical Trials of PDD

Conclusions

Early clinical trials of PDD have demonstrated significant toxic side effects and modest therapeutic benefit.

The limiting toxicity appears to be cumulative renal functional impairment and apparently irreversible high frequency hearing loss. Myelosuppression appears to be reversible. Toxicity is roughly dose related. No idiosyncratic reactions were observed.

Of 36 evaluable patients, ten experienced minor therapeutic effect. It is of interest that most of these responses were observed in patients with epidermoid carcinomas of the head and neck. Although the responses were limited in degree and duration, they demonstrated the definite anti-tumor potential of PDD. Further studies of PDD in combination with other agents are being undertaken [9]. We are uncertain of the future of PDD in cancer chemotherapy. The reports of new analogs with possibly improved therapeutic index may lead to other, related, more effective agents.

References

1. Broder, L., Carter, S.: Clinical brochure: *Cis*-platinum(II) diamminedichloride. National Cancer Institute Chemotherapy Branch, June 1971.
2. Harder, H., Rosenberg, B.: Inhibitory effects of antitumor platinum compounds on DNA, RNA and protein synthesis in mammalian cells *in vitro*. Int. J. Cancer **6**, 207 (1970).

3. Hill, J., Speer- R., Loeb, E., MacLellan, A., Hill, N., Khan, A.: Clinical experience with *cis*-platinous diamminedichloride, advances in antimicrobial and antineoplastic chemotherapy. Vol. 2, p. 255. München, Berlin, Wien: Urban and Schwarzenberg 1972.
4. Howle, J., Gale, G.: *Cis*-dichlorodiammine-platinum(II). Cytological changes induced in *E. coli.* J. Bact. **103,** 258 (1970).
5. Howle, J., Gale, G.: *Cis*-dichlorodiammine-platinum(II). Persistant and selective inhibition of deoxyribonucleic acid synthesis *in vivo.* Biochem. Pharmacol. **19,** 2757 (1970).
6. Howle, J., Thompson, H., Stone, A., Gale, G.: Cis-dichlorodiammine-platinum (II): Inhibition of nucleic acid synthesis in lymphocytes stimulated with phytohemagglutinin. Proc. Soc. exp. Biol. (N. Y.) **137,** 820 (1971).
7. Karnofsky, D. A.: Meaningful clinical classification of therapeutic responses to anticancer drugs. Clin. Pharmacol. Ther. **2,** 709 (1961).
8. Kociba, R., Sleight, S., Rosenberg, B.: Inhibition of dunning ascitic leukemia and Walker 256 carcinosarcoma with *cis*-diamminedichloro-platinum (II) (NSC-119 875). Cancer Chemother. Reports (Part 1) **54,** 325 (1970).
9. Lippman, A., Helson, C., Helson, L., Kaufman, R., Krakoff, I. H.: Clinical trials of *cis*-platinum (II) diamminedichloride (PDD). Proc. Amer. Ass. Cancer Res. **13,** 157 (1972).
10. Lippman, A. J., Helson, C., Helson, L., Krakoff, I. H.: Clinical trials of *cis*-platinum-(II) diamminedichloride. Cancer Chemother. Rep. **57,** 191 (1973).
11. Rosenberg, B., van Camp, L.: The successful regression of large solid sarcoma 180 tumors by platinum compounds. Cancer Res. **30,** 1799 (1970).
12. Rosenberg, B., van Camp, L., Grimley, E., Thomson, A.: The inhibition of growth or cell division in *E. coli* by different species of platinum(IV) complexes. J. biol. Chem. **242,** 1347 (1967).
13. Rosenberg, B., van Camp, L., Krigas, T.: Inhibition of cell division in *E. coli* by electrolysis products from a platinum electrode. Nature (Lond.) **205,** 698 (1965).
14. Rosenberg, B., van Camp, L., Trosko, J., Mansour, V.: Platinum compounds: a new class of potent antitumor agents. Nature (Lond.) **222,** 385 (1969).
15. Thompson, H., Gale, G.: *Cis*-dichlorodiammine-platinum (II): Hematopoietic effects in rats. Toxicol. appl. Pharmacol. **19,** 602 (1971).
16. Welsch, C.: Effects of *cis*-platinum (II) diamminedichloride on growth of 7,12-dimethylbenzanthracene (DMBA) induced mammary tumors in female rats. Proc. Amer. Ass. Cancer Res. **12,** 25 (1971).
17. Zak, M., Drobnik, S., Rezny, Z.: The effects of *cis*-platinum (II) diamminedichloride on bone marrow. Cancer Res. **32,** 595 (1972).

Preliminary Experience with *cis*-Platinum (II) Diamminedichloride (PDD)

Z. Dienstbier, O. Andrysek, and J. Zamecnik

Several publications from this Institute deal with the additive toxicity of *cis*-platinum(II) diamminedichloride (PDD) and X-irradiation and with the quantitative changes in bone marrow induced by PDD [10, 12].

We have shown decreased mitotic activity and cellularity of bone marrow after administration of PDD and, unlike other cytostatics, a rapid regeneration. Fremuth *et al.* [3] have studied chromosomal aberrations in the Chinese hamster and have shown that in bone marrow, PDD causes more breakages than irradiation.

Table 1. Treated neoplasma

Types of neoplasma	Dosage	Drug toxicity	Obj. regressions
1. Ca colon general.	10 mg/m^2 5×	anemia; $>$ Le in urine $>$ SGPT; $>$ SGOT	none
2. Ca pancreas general.	10 mg/m^2 5×	anemia; $>$ SGPT; $>$ SGOT	none
3. Melanoma general.	10 mg/m^2 7×	anemia; thrombocytopenia; $>$ Le in urine	none
4. Retothelosarcoma gen.	10 mg/m^2 5×	—	none
5. Ca hepar meta	20 mg/m^2 6×	$>$ Le+cylinders in urine $<$ renal filtration	scintigraphic regression of liver tumor

In clinical studies, PDD has so far been administered to 5 patients. Table 1 summarises the diagnoses and dose levels used. In agreement with Rossof *et al.* [6] and Krakoff [4] it was found that the toxicity of PDD was largely reversible. The most common side effect was vomiting, which was suppressed by Torecan (Sandoz).

In one patient a clear effect of PDD was observed. The patient KK (female aged 65) had been found to have a tumour of the liver during explorative laparotomy. The tumour was diagnosed as a necrotic tubular adenomatous carcinoma of unknown primary origin. The gammagram of liver before and after PDD is shown in Figs. 1 and 2. The patient, who had not been previously treated, received 20 mg/m^2 of PDD

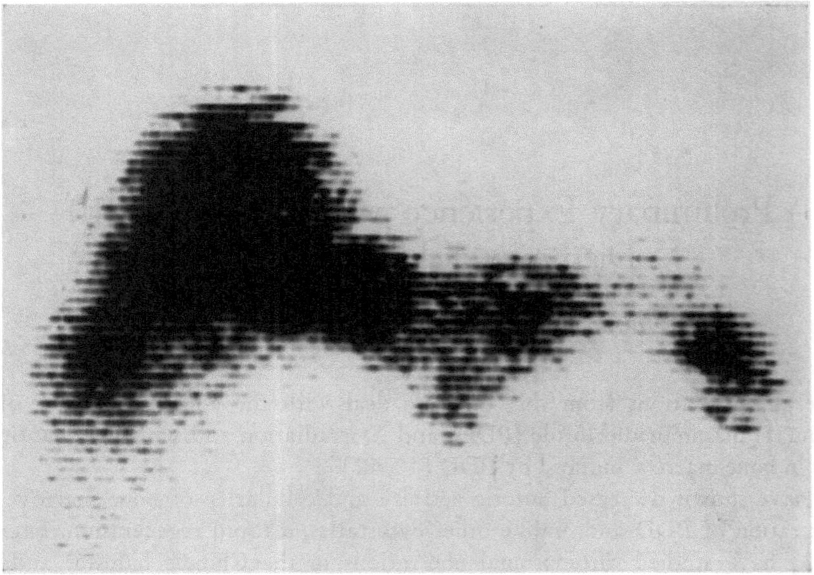

Fig. 1. Gammagram of liver of patient with metastatic hepatic carcinoma. Before treatment

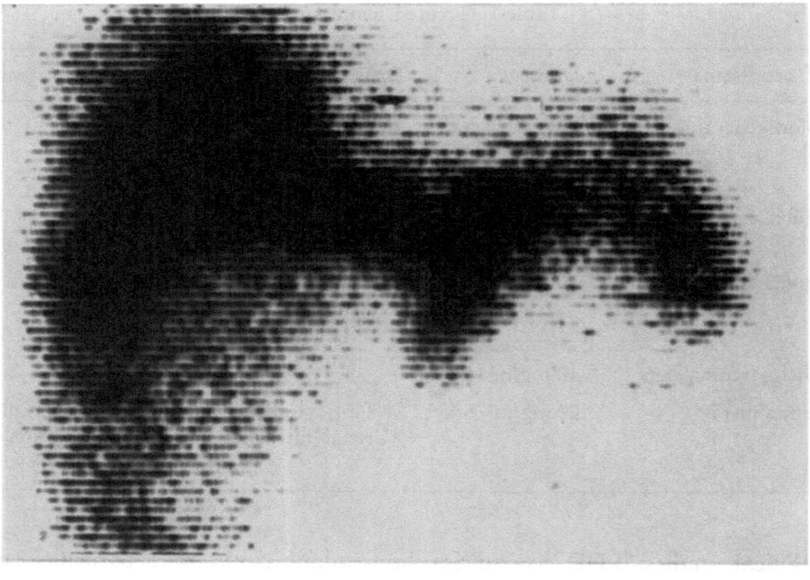

Fig. 2. Gammagram of liver of patient two weeks after the end of PDD therapy (20 mg/m^2$\times$6 at weekly intervals)

in six injections, separated by weekly intervals. No significant change occurred in the blood picture. Blood urea nitrogen was not increased in spite of decreased filtration and reabsorption by the kidney. However, in the Addis sediment, leucocytes were found in an increased amount and cylinders were also observed in the urine after a week. The patient is under constant medical supervision.

In spite of some effects in our patients we are not optimistic about the use of platinum complexes as single agents and confirm the findings of Rossof *et al.* [6], Perlia [5], Talley [8], Wallace [9] and Steuart [7] that the therapeutic action of PDD is slight.

In addition to normal laboratory tests the effect of PDD on the excretion of pyrimidine nucleosides in the urine is also being examined. Unlike irradiation and nitrogen mustard no increased excretion of pyrimidine nucleosides occurred in rats up to 24 h after administration of PDD [2].

In man it has been established that deoxycytidine, deoxyuridine and thymidine are excreted in a dose dependent manner following administration of a cytostatic agent [1]. Similar research in man with the platinum complexes may help in the elucidation of their mechanism of action.

References

1. Dienstbier, Z., Bechyně, M., Burič, L., Kovaříková, H., Blažek, L.: Clinical use of deoxypyrimidinuria in the course of the therapy of Hodgkin's disease. Čas. Lék. čes. **112**, 475 (1973).
2. Dienstbier, Z., Blažek, L.: Unpublished preliminary results 1973.
3. Fremuth, F., Dienstbier, Z., Bárta, I., Drobník, J., Bradna, F.: Effect of radiation, chemomutagenes and radioprotective compounds on the frequency of intergene mutations in bone marrow of Chinese hamsters. Book of Abstracts. VIIth International Congress of Chemotherapy, vol. II-C-1/5, 1971.
4. Krakoff, I.: *Cis*-platinum (II) diamminedichloride. Sloan-Kettering Studies 1972.
5. Perlia, Ch.: *Cis*-platinum (II) diamminedichloride. Eastern Cooperative Oncology Group Studies 1972.
6. Rossof, A. H., Slayton, R. E., Perlia, Ch. P.: Preliminary clinical experience with *cis*-diamminedichloro-platinum (II) (NSC-119 875, CACP). Reprint VIII. Annual Session of ASCO, Boston 1972.
7. Steuart, Ch.: *Cis*-platinum (II) diamminedichloride. Southeastern Cooperative Group Studies 1972.
8. Talley, R.: *Cis*-platinum (II) diamminedichloride. Southwest Group Studies 1972.
9. Wallace, H. J.: *Cis*-platinum (II) diamminedichloride. Rosswell Parl Studies 1972.
10. Žák, M., Drobník, J.: Effect of *cis*-dichlorediammine-platinum (II) on the post-irradiation lethality in mice after irradiation with X-rays. Strahlentherapie **142**, 115 (1971).
11. Žák, M., Drobník, J., Režny, Z.: Effect of *cis*-dichlorediammino-platinum (II) on cytological and morphological changes in bone marrow of mice (long term observations). Neoplasma **19**, 305 (1972).
12. Žák, M., Drobník, J., Režny, Z.: The effect of *cis*-platinum (II) diamminedichloride on bone marrow. Cancer Res. **32**, 595 (1972).

Clinical Experience with *cis*-Platinum (II) Diamminedichloride

I. Hindy, C. Sellei, R. Varsanyi, and S. Eckhardt

From the experimental data of Rosenberg *et al.* [3, 4] and others [1, 2] it is clear that various platinum complexes have good antitumour properties. Their primary action probably involves inhibition of DNA synthesis. At the National Oncological Institute in Budapest preclinical investigations of *cis*-platinum(II) diamminedichloride (PDD) have been carried out by Kellner, Nemeth and Somfai. At 3×2.5 mg/kg and 3×1.25 mg/kg of PDD respectively 66% and 88% of mice bearing the NK/Ly ascites were free from tumour at 101 days. 67—76% inhibition of the sarcoma 180 was also obtained although only at dose levels causing a considerable

Table 1. The data of patients treated by PDD

Name	Sex	Age	Diagnosis	Dose
B. J.	♂	63	Tu. parotis	1×70 mg
R. L.	♀	54	Tu. recti	1×50 mg
S. J.	♀	47	Tu. mammae	4×18 mg
B. M.	♀	73	Tu. coeci	4×12 mg
K. B.	♂	71	Tu. parotis	4×18 mg
H. J.	♂	40	Tu. epiglottis	4×12 mg 1×10 mg
N. Gy.	♂	43	Tu. met. colli	4×15 mg
M. L.	♀	74	Tu. mammae	1×10 mg

diminuition of spleen weight indicating drug toxicity. In a short time course study using the Guerin tumour maximum effects were seen 48 h after drug administration and consisted of abnormal mitoses and giant cells.

Table 1 summarises the patients who have so far been treated with PDD. All patients (4 male, 4 female from 43—74 years of age) had solid tumours and had previously received chemotherapy, X-irradiation or both. In patient no. 7, the site of the primary tumour was not identified, the metastatic tumour on the neck being a squamous cell cancer.

The first 2 patients received 1 injection of 1 mg PDD/kg. No objective remissions were observed although the side effects were marked, especially the vomiting. The other patients were therefore given the drug at 0.25 mg/kg on four consecutive days. Patient no. 6 was given a second treatment, 4 weeks after the first, but it was discontinued due to his generally poor condition. The last patient has also only received a single dose of 0.25 mg/kg of PDD since her haematological picture was not satisfactory. Some objective tumour reduction was observed in patients receiving the 4 daily doses of PDD and some subjective improvement of their general condition. All effects however, were transitory not lasting more than two to 3 weeks.

Particular attention was paid to kidney and liver function since heavy metals can interfere with these functions, but no detrimental effects were observed. In one case there was an elevation of the level of BUN and in another case cylinders in the urine but in both cases patients were normal three to four days later. All patients had a complete haematological examination before and after the treatment; a transitory decrease in platelet count (2 cases) and leukopenia (1 case) were the only findings. Vomiting was the most frequent side effect and was observed not only in the first two patients, who received a large single dose of PDD, but in almost every case, even where the drug was administered in 500 ml of saline.

No patients died within six weeks of treatment. No kidney or liver damage related to the PDD treatment could be established post mortem.

References

1. Gale, G. R., Howle, J., Walker, E. M.: Anti-tumour and anti-mitogen properties of *cis*-dichloro(dipyridine)platinum(II). Cancer Res. **31**, 950 (1971).
2. Leonard, B. J., Eccleston, E., Jones, D., Todd, P., Walpole, A.: Anti-leukaemic and nephrotoxic properties of platinum compounds. Nature (Lond.) **234**, 43 (1971).
3. Rosenberg, B., van Camp, L.: The successful regression of large solid sarcoma 180 tumors by platinum compounds. Cancer Res. **30**, 1799 (1970).
4. Rosenberg, B., van Camp, L., Trosko, J. E., Mansour, V. H.: Platinum compounds: a new class of potent antitumour agent. Nature (Lond.) **222**, 385 (1969).

Subject Index

Recent Results in Cancer Research

Sponsored by the Swiss League against Cancer. Editor in Chief: P. Rentchnick, Genève

1 SCHINDLER, R.: Die tierische Zelle in Zellkultur. Geb. DM 19,80; US $ 8.10
2 Neuroblastomas — Biochemical Studies. Edited by C. BOHUON, (Symposium). Cloth DM 19,80; US $ 8.10
3 HUEPER, W. C.: Occupational and Environmental Cancers of the Respiratory System. Cloth DM 41,00; US $ 16.80
4 GOLDMAN, L.: Laser Cancer Research. Cloth DM 19,80; US $ 8.10
5 METCALF, D.: The Thymus. Its Role in Immune Responses, Leukaemia Development and Carcinogenesis. Cloth DM 29,00; US $ 11.90
6 Malignant Transformation by Viruses. Edited by W. H. KIRSTEN, (Symposium). Cloth DM 39,00; US $ 16.00
7 MOERTEL, CH. G.: Multiple Primary Malignant Neoplasms. Their Incidence and Significance. Cloth DM 22,00; US $ 9.00
8 New Trends in the Treatment of Cancer. Edited by L. MANUILA, S. MOLES, and P. RENTCHNICK. Cloth DM 39,00; US $ 16.00
9 LINDENMANN, J., and P. A. KLEIN: Immunological Aspects of Viral Oncolysis. Cloth DM 22,00; US $ 9.00
10 NELSON, R. S.: Radioactive Phosphorus in the Diagnosis of Gastrointestinal Cancer. Cloth DM 19,00; US $ 7.80
11 FREEMAN, R. G., and J. M. KNOX: Treatment of Skin Cancer. Cloth DM 19,00; US $ 7.80
12 LYNCH, H. T.: Hereditary Factors in Carcinoma. Cloth DM 29,00; US $ 11.90
13 Tumours in Children. Edited by H. B. MARSDEN and J. K. STEWARD. Cloth DM 87,00; US $ 35.50
14 ODARTCHENKO, N.: Production Cellulaire Erythropoiétique. Relié DM 34,00; US $ 13.90
15 SOKOLOFF, B.: Carcinoid and Serotonin. Cloth DM 29,00; US $ 11.90
16 JACOBS, M. L.: Malignant Lymphomas and Their Management. Cloth DM 22,00; US $ 9.00
17 Normal and Malignant Cell Growth. Edited by R. J. M. FRY, M. L. GRIEM, and W. H. KIRSTEN (Symposium). Cloth DM 63,00; US $ 25.70
18 ANGLESIO, E.: The Treatment of Hodgkin's Disease. Cloth DM 26,50; US $ 10.90
19 BANNASCH, P.: The Cytoplasm of Hepatocytes during Carcinogenesis. Electron- and Lightmicroscopical Investigations of the Nitrosomorpholine-intoxicated Rat Liver. Cloth DM 35,50; US $ 14.50
20 Rubidomycin. A new Agent against Cancer. Edited by J. BERNARD, R. PAUL, M. BOIRON, C. JACQUILLAT, and R. MARAL. Cloth DM 53,00; US $ 21.70
21 Scientific Basis of Cancer Chemotherapy. Edited by G. MATHÉ (Symposium). Cloth DM 31,00; US $ 12.70
22 KOLDOVSKÝ, P.: Tumor Specific Transplantation Antigen. Cloth DM 26,50; US $ 10.90
23 FUCHS, W. A., J. W. DAVIDSON, and H. W. FISCHER: Lymphography in Cancer. With contributions by G. JANTET and H. RÖSLER. Cloth DM 84,00; US $ 34.30
24 HAYWARD, J. L.: Hormones and Human Breast Cancer. An Account of 15 Years Study. Cloth DM 37,50; US $ 15.30
25 ROY-BURMAN, P.: Analogues of Nucleic Acid Components. Mechanisms of Action. Cloth DM 31,00; US $ 12.70
26 Tumors of the Liver. Edited by G. T. PACK and A. H. ISLAMI. Cloth DM 62,00; US $ 25.30
27 SZYMENDERA, J.: Bone Mineral Metabolism in Cancer. Cloth DM 35,50; US $ 14.50
28 MEEK, E. S.: Antitumour and Antiviral Substances of Natural Origin. Cloth DM 18,00; US $ 7.40
29 Aseptic Environments and Cancer Treatment. Edited by G. MATHÉ (Symposium). Cloth DM 24,50; US $ 10.00
30 Advances in the Treatment of Acute (Blastic) Leukemias. Edited by G. MATHÉ (Symposium). Cloth DM 42,00; US $ 17.20
31 DENOIX, P.: Treatment of Malignant Breast Tumors. Indications and Results. Cloth DM 53,00; US $ 21.70

32 NELSON, R. S.: Endoscopy in Gastric Cancer. Cloth DM 53,00; US $ 21.70

33 Experimental and Clinical Effects of L-Asparaginase. Edited by E. GRUNDMANN and H. F. OETTGEN (Symposium). Cloth DM 64,00; US $ 26.20

34 Chemistry and Biological Actions of 4-Nitroquinoline 1-Oxide. Edited by H. ENDO, T. ONO, and T. SUGIMURA. Cloth DM 36,—; US $ 14.70

35 PENN, I.: Malignant Tumors in Organ Transplant Recipients. Cloth DM 26,50; US $ 10.90

36 Current Concepts in the Management of Lymphoma and Leukemia. Edited by J. E. ULTMANN, M. L. GRIEM, W. H. KIRSTEN, and R. W. WISSLER (Symposium). Cloth DM 48,—; US $ 19.60

37 CHIAPPA, S., R. MUSUMECI, and C. USLENGHI: Endolymphatic Radiotherapy in Malignant Lymphomas. With contributions by G. BONADONNA, B. DAMASCELLI, G. FAVA, F. PIZZETTI, U. VERONESI. Cloth DM 48,—; US $ 19.60

38 KOLLER, P. C.: The Role of Chromosomes in Cancer Biology. Cloth DM 48,—; US $ 19.60

39 Current Problems in the Epidemiology of Cancer and Lymphomas. Edited by E. GRUNDMANN and H. TULINIUS (Symposium). Cloth DM 58,—; US $ 23.70

40 LANGLEY, F. A., A. C. CROMPTON: Epithelial Abnormalities of the Cervix Uteri. Cloth DM 58,—; US $ 23.70

41 Tumours in a Tropical Country. A Survey of Uganda (1964—1968). Edited by A. C. TEMPLETON. Cloth DM 72,—; US $ 29.40

42 Breast Cancer: A Challenging Problem. Edited by M. L. GRIEM, E. V. JENSEN, J. E. ULTMANN, and R. W. WISSLER (Symposium). Cloth DM 48,—; US $ 19.60

43 Nomenclature, Methodology and Results of Clinical Trials in Acute Leukemias. Edited by G. MATHÉ, P. POUILLART, and L. SCHWARZENBERG (Symposium). Cloth DM 58,—; US $ 23.70

44 Special Topics in Carcinogenesis. Edited by E. GRUNDMANN (Symposium). Cloth DM 58,—; US $ 23.70

45 KOLDOVSKÝ, P.: Carcinoembryonic Antigens. Cloth DM 38,—; US $ 15.50

46 Diagnosis and Therapy of Malignant Lymphoma. Edited by K. MUSSHOFF (Symposium). Cloth DM 62,—; US $ 25.30

47 Investigation and Stimulation of Immunity in Cancer Patients. Edited by G. MATHÉ and R. WEINER (Symposium). Cloth DM 96,—; US $ 39.20

48 Platinum Coordination Complexes in Cancer Chemotherapy. Edited by T. A. CONNORS and J. J. ROBERTS (Symposium). Cloth DM 68,—; US $ 27.80

49 Complications of Cancer Chemotherapy. Edited by G. MATHÉ and R. K. OLDHAM (Symposium). Cloth DM 58,—; US $ 23.70

Special Supplement: Biology of Amphibian Tumors. Edited by M. MIZELL. Cloth DM 96,00; US $ 39.20

In Production

50 Cancer Registry. Edited by E. GRUNDMANN and E. PEDERSEN (Symposium)

51 Cancer Chemotherapy. Edited by A. CLARYSSE, E. KENIS, and G. MATHÉ

Prices are Subject to change without notice

MIX
Papier aus verantwortungsvollen Quellen
Paper from responsible sources
FSC® C105338

If you have any concerns about our products,
you can contact us on
ProductSafety@springernature.com

In case Publisher is established outside the EU,
the EU authorized representative is:
Springer Nature Customer Service Center GmbH
Europaplatz 3, 69115 Heidelberg, Germany

Printed by Libri Plureos GmbH
in Hamburg, Germany